Electronic Noses and Olfaction 2000

T0203715

Sensors Series
Series Editor: **B E Jones**

Solid State Gas Sensors
Edited by P T Moseley and B C Tofield

Techniques and Mechanisms in Gas Sensing
Edited by P T Moseley, J O W Norris and D E Williams

Hall Effect Devices
R S Popović

Thin Film Resistive Sensors
Edited by P Ciureanu and S Middelhoek

Biosensors: Microelectrochemical Devices
M Lambrecht and W Sansen

Sensors VI: Technology, Systems and Applications
Edited by K T V Grattan and A T Augousti

Automotive Sensors
M H Westbrook and J D Turner

Sensors and their Applications VII
Edited by A T Augousti

Advances in Actuators
Edited by A P Dorey and J H Moore

Intelligent Sensor Systems, Revised Edition
J E Brignell and N M White

Sensor Materials
P T Moseley and J Crocker

Ultrasonic Sensors
R C Asher

Sensors and their Applications VIII
Edited by A T Augousti and N M White

Eurosensors XII
Edited by N M White

Sensors and their Applications X
Edited by N M White and A T Augousti

Electronic Noses and Olfaction 2000

**Proceedings of the Seventh International Symposium
on Olfaction and Electronic Noses,
held in Brighton, UK, July 2000**

Edited by

J W Gardner
University of Warwick

and

Krishna C Persaud
UMIST

CRC Press
Taylor & Francis Group
Boca Raton London New York

CRC Press is an imprint of the
Taylor & Francis Group, an **informa** business

CRC Press
Taylor & Francis Group
6000 Broken Sound Parkway NW, Suite 300
Boca Raton, FL 33487-2742

First issued in paperback 2019

© 2000 by Taylor & Francis Group, LLC
CRC Press is an imprint of Taylor & Francis Group, an Informa business

No claim to original U.S. Government works

ISBN-13: 978-0-367-39766-1

**Visit the Taylor & Francis Web site at
http://www.taylorandfrancis.com**

**and the CRC Press Web site at
http://www.crcpress.com**

British Library Cataloguing in Publication Data

A catalogue record for this book is available from the British Library.

Library of Congress Cataloging-in-Publication Data are available

Series Editor: **Professor B E Jones**, Brunel University

Preface

This book reflects the current state of progress towards the development and application of electronic instruments called 'electronic noses' or, more recently, 'e-noses'. The instruments are generally based on arrays of sensors for volatile chemicals with broadly tuned selectivity, coupled to appropriate pattern recognition systems. They are capable of detecting and discriminating a number of different simple and complex odours, such as the headspace of coffee and olive oil, as well as being able to perform simple multi-component gas analysis.

This book contains a series of papers based on material that was originally presented at the 7th international symposium on olfaction and electronic noses (ISOEN 2000) held in Brighton, UK, July 2000. This series of symposia was started up in 1994 by the e-nose manufacturer Alpha MOS (France) to promote the field and further our understanding of this complex subject. The Brighton symposium was attended by scientists, engineers, technologists, clinicians and investigators, as well as instrument manufacturers, from Asia, Europe and the USA — many of them working in the diverse areas of applied research on e-noses — as well as those interested in the plethora of applications in olfaction and taste.

The contributions covered a mixture of research and state-of-the-art papers on numerous aspects of e-noses such as sensors, materials, sampling methods, instrumentation and pattern recognition techniques, together with selected applications. Consequently, this book provides a valuable up-date of the latest developments in the field of electronic noses.

J W Gardner
University of Warwick

K C Persaud
UMIST

Contents

Section 1

Odours, Taste and Physico-chemical Interactions

Section 1: Odours, Taste and Physico-chemical Interactions
Paper presented at the Seventh International Symposium on Olfaction and Electronic Noses, July 2000

3

Oral Malodour

Sushma Nachnani MS

University Health Resources Group, Inc
Email: sushman@worldnet.att.net

Abstract

Oral malodour has been recognized in the literature since ancient times, but in the last five to six years it has increasingly come to the forefront of public and dental professional awareness. Approximately 40 –50 % of dentists see 6-7 self- proclaimed oral malodour patients per week. Oral malodour research has gained momentum with increasing suspicions being directed at the sulfur-producing bacteria as the primary source of this condition. This chapter outlines a basic overview of oral malodour condition and provides a detailed discussion of most of the antimicrobial agents presently involved or suggested in the treatment of oral malodour Oral malodour can be caused by many localized and systemic disorders. Oral Malodour (OM) caused by normal physiological processes is usually transitory. Chronic or pathological halitosis stems from oral or non-oral sources. In the later case we call this condition oral malodour. Some of the oral causes are periodontal disease, gingivitis, and plaque coating on the dorsum of the tongue. Nasal problems such as postnasal drip that falls at the posterior dorsum of the tongue may exacerbate the oral malodour condition. A relationship between gastrointestinal diseases such as gastritis and oral malodour has not been established. Oral malodour has its basis due to bacterial putrefaction. Many patients with a chief complaint of oral malodour have some level of gingival and or periodontal pathology sufficient to be the etiology, but clearly periodontol pathology is not a prerequisite for production of oral malodour. OM in healthy patients arises from the oral cavity and generally originates on the dorsum of the tongue. Cadaverine levels have been reported to be associated with oral malodour and this association may be independent of VSC. Subjects without the history of oral malodour that rinsed with cysteine produced high oral concentrations of VSC, suggesting that cysteine is a major substrate for VSC production.

Introduction

Oral malodour has been recognized in the literature since ancient times, but in the last five to six years it has increasingly come to the forefront of public and dental professional awareness [1]. Approximately 40 –50 % of dentists see 6-7 self- proclaimed oral malodour patients per week [2]. Standard diagnosis and treatment for oral malodour in the routine care of each patient has not been established in the dental or the medical field. However the transfer of knowledge is increasing because of pioneering researchers and clinicians that have developed reputable clinics dealing with this condition. Dental and medical schools must incorporate diagnosis and treatment of oral malodour in their curriculum, so that the future generations of clinicians can effectively treat this condition. Although this area of research has been ridiculed, at least 50 % of the population [3] suffers from a chronic oral malodour condition by which individuals experience personal discomfort and social embarrassment leading to emotional distress. The consequences of oral malodour may be more than social; it may reflect serious local or systemic conditions. Oral malodour research has gained momentum with increasing suspicions being directed at the sulfur-producing bacteria as the primary source of this condition. This chapter outlines a basic overview of oral malodour condition and provides a detailed discussion of most of the antimicrobial agents presently involved or suggested in the treatment of oral malodour.

Oral and non-oral causes

Oral malodour can be caused by many localized and systemic disorders. Oral Malodour (OM) caused by normal physiological processes is usually transitory. Transient or non pathologic OM may due to hunger, low levels of salivation during sleep food debris, prescription of drugs and smoking [4]. Chronic or pathological halitosis stems from oral or non- oral sources. The mucosa of the mouth and the upper respiratory tract are used to expel volatile compounds from the body. These can be gases metabolic end products produced by diet such as garlic, alcohol etc. . These volatile compounds may be produced in extra oral cavities as well as in the oral cavity. In the later case we call this condition oral malodour. In addition there appear to be several other metabolic conditions involving enzymatic and transport anomalies (such as Trimethylaminuria) which lead to systemic production of volatile malodours that manifest themselves as halitosis and/or altered chemoreception[5]. Some of the oral causes are periodontal disease, gingivitis, and plaque coating on the dorsum of the tongue. OM may be aggravated by a reduction in salivary flow. Radiation therapy, Sjogren's Syndrome, lung cancer, peritonsillar abscess, cancer of the pharynx and cryptic tonsils can also contribute to oral malodour [5]. Nasal problems such as postnasal drip that falls at the posterior dorsum of the tongue may exacerbate the oral malodour condition. Odour generated in this manner can be easily distinguished from mouth odour by comparing the odour exiting the mouth or nose [6]. The non-oral causes of OM include diabetic ketosis, uremia, gastrointestinal conditions, and irregular bowel movement, hepatic and renal failure and certain types of carcinomas such as leukemia. The accurate clinical labeling and interpretation of different oral malodours both contribute to the diagnosis and treatment of underlying disease [7] (Table 9-1). Taste and smell can be altered due to facial injuries, cosmetic surgery radiation and olfactory epithilium located on the dorsal aspect of the nose [8]. A relationship between gastrointestinal diseases such as gastritis and oral malodour has not been established. However, oral malodour has been reported in some patients with a history of gastritis, or duodenal and gastric ulcers [10].

Oral malodour has its basis due to bacterial putrefaction, the degradation of proteins, and the resulting amino acids produced by microorganisms [11]. Many patients with a chief complaint of oral malodour have some level of gingival and or periodontal pathology sufficient to be the etiology, but clearly periodontol pathology is not a prerequisite for production of oral malodour [12]. Medications such as antimicrobial agents, antirheumatic, anti hypertentive, antidepressants and analgesics may cause altered taste and xerostomia.

OM in healthy patients arises from the oral cavity and generally originates on the dorsum of the tongue [6][13][14][15][16]. The volatile sulfur compound (VSC) producing anaerobic bacteria appear to be the primary source of this odours[16] The bacteria hydrolyze the proteins to amino acids, three of which contain sulfur groups and are the precursors to volatile sulfur compounds (VSCs). These gaseous substances, responsible for malodour, consist primarily of hydrogen sulfide (H_2S), dimethyl sulfide [$(CH_3)_2S$], methyl mercaptan (CH_3SH) and sulfur dioxide (SO_2)[11][13][15][16]. Cadaverine levels have been reported to be associated with oral malodour and this association may be independent of VSC [17]. Subjects without the history of oral malodour that rinsed with cysteine produced high oral concentrations of VSC, suggesting that cysteine is a major substrate for VSC production. The other sulfur-containing amino acids had much less effect on VSC production. It was found that the tongue was the major site for VSC production [18]

The tongue-coating of the plaque

Research suggests that the tongue is the primary site in the production of OM. The dorsoposterior surface of the tongue has been identified as the principal location of the intraoral generation of VSC's [20]. The tongue is an excellent site for the growth of microorganisms since the papillary nature of the tongue dorsum creates a unique ecological site that provides an extremely large surface area, favoring the accumulation of oral bacteria. The proteolytic, anaerobic bacteria that reside on the tongue play an essential part in the development of oral malodour. The presence of tongue coating has been shown to have a correlation with the density or total number of bacteria in the tongue plaque coating [21]. The weight of the tongue coating in periodontal patients was elevated to 90mgs, while the VSC was increased by a factor of 4 and CH_3SH/H_2S ratio was increased 30-fold when compared with individuals with healthy periodontium[20]. This high ratio of amino acids can be due to free amino acids in the cervicular fluid when compared with those of L- cysteine [20]. The BANA (Benzoyl –DL-arginine-2 napthylamide) test has been used to detect *Treponema denticola* and *Porphyromonas gingivalis* and *Bacteroides forsythus*. The three organisms that may contribute to oral malodour can be easily detected by their capacity to hydrolyze BANA, a trypsin-like substrate. BANA scores are associated with a component of oral malodour, which is independent of volatile sulfide measurements, and suggest its use as an adjunct test to volatile sulfide measurement [22]. Higher mouth odour organoleptic scores are associated with heavy tongue coating and correlate with the bacterial density on the tongue and it also correlates to BANA-hydrolyzing bacteria -*T.denticola, P gingivalis, and B. forsythus* [23]

Microbiota associated with oral malodour

The actual bacterial species that cause OM have yet to be identified from among the 300 plus bacterial species in the mouth. Putrefaction is thought to occur under anaerobic conditions, involving a range of bacteria such as *Fusobacterium, Veillonella, T. denticola, P. gingivalis, Bacteroides and Peptostreptococcus* [23][24]. Studies have shown that essentially all odour production is a result of gram-negative bacterial metabolism and that the gram-positive bacteria contribute very little odour [24]. *Fusobacterium nucleatum* is one of the predominant organisms associated with gingivitis and periodontitis and this organism produces high levels of VSCs. The nutrients for the bacteria are provided by oral fluids, tissue and food debris. Methionine is reduced to methyl mercaptan and cysteine. Cysteine is reduced to cystine, which is further reduced to hydrogen sulfide in the presence of sulfhydrase-positive microbes. This activity is favored at a pH of 7.2 and inhibited at a pH of 6.5 [11][13][15] [16]. Isolates of *Klebsiella* and *Enterobacter* emitted foul odours in vitro which resembled bad breath with concomitant production of volatile sulfides and cadaverine both compounds related to bad breath in denture wearers [26].The amounts of volatile sulfur compounds (VSC) and methyl mercaptan/hydrogen sulfide ratio in mouth air from patients with periodontal infection were reported to be eight times greater than those of control subjects [16].

Oral malodour assessment parameters

Organoleptic Measurements - One major research problem that must be tackled is the lack of an established "gold standard" for rapidly measuring OM condition. The objective assessment of oral malodour is still best performed by the human sense of smell (direct sniffing-organoleptic method) but more quantifiable measures are being developed. At present, confidant feedback and expert odour (organoleptic) judges are the most commonly used approaches. Both assessments use a 0-5 scale in order to consistently quantify the odour (0= No odour present, 1= Barely noticeable odour, 2= Slight but clearly noticeable odour, 3= Moderate odour, 4= Strong offensive odour, 5= Extremely foul odour). Individuals are instructed to refrain from using any dental products, eating or using deodorants of fragrances four hours prior to the visit to the clinic Individuals are also advised to bring their confidante or friends to assess their oral malodour. In order to create a reproducible assessment, subjects are instructed to close their mouth for two minutes and not to swallow during that period. After two minutes the subject breathes out gently, at a distance of 10 cm from the nose of the their counterpart and the organoleptic odours are assessed [27] In order to reduce inter-examiner variations, a panel consisting of several experienced judges is often employed. A study on the inter-examiner reproducibility indicates that there is some co-relation, albeit poor [28]. Gender and age influence the performance of an organoleptic judge. Females have a better olfactory sense and it decreases with age. Dentists and periodontist may not be ideal judges if they do not use masks on a daily basis [29].

OM can be analyzed using gas chromatography (GC) coupled with flame photometric detection [30] This allows separation and quantitative measurements of the individual gasses. However the equipment necessary is expensive and requires skilled personnel to operate it. This equipment is also cumbersome and the analysis is time consuming. As a result, GC cannot be used in the dental office and is not always used in OM clinical trials.
The portable sulfide meter (Halimeter Interscan, Inc Chatsworth CA.) has been widely used over the last few years in OM testing. The portable sulfide meter is an electrochemical, voltmetric sensor which generates a signal when it is exposed to sulfide and mercaptan gases and measures the concentration of hydrogen sulfide gas in parts per billion. The halimeter is portable and does not require skilled personnel for operation. The main disadvantages of using this instrument are the necessity of periodic re-calibration and the measurements cannot be made in the presence of ethanol or essential oils [28] In other words, the measurements may be affected if the subject is wearing perfume, hairspray, deodorant, etc. In addition, this limitation does not allow the assessment of mouthwash efficacy until after these components have dissipated.

The Electronic Nose

The " Electronic Nose" is a hand held device, being developed to rapidly classify the chemicals in unidentified vapor. It's application by scientists and personnel in the medical and dental field as well as it is hoped that this technology will be inexpensive, miniaturizable and adaptable to practically any odour detecting task [31] If the "Electronic Nose" can learn to "smell" in a quantifiable and reproducible manner, this tool will be a revolutionary assessment technique in the field of OM This device is based on sensor technology that can " smell" and produce unique fingerprints for distinct odours Preliminary data indicates that this device has a potential to be used as a diagnostic tool to detect odours.

Management of oral malodour

Large number of so called " Fresh Breath Clinics" are offering diagnostic and treatment services for patient's complaints of " oral malodour". There are no accepted standards of care for these services, and the clinical protocols vary widely [34][35].
A thorough medical, dental and halitosis history is necessary to determine whether the patient's complaint of bad breath is due to oral causes or not [32]. It is important to determine the source of oral malodour; complaints about bad taste should be noted. In most cases patients that complain of bad taste may not have bad breath. The taste disorders may be due to other causes[32]. It has been reported that in approximately 8% of the individuals the odour was caused by tonsilitis, sinusitis, or a foreign body in the nose [35] . Approximately 80-90% of the oral malodour originates from the dorsum of the tongue. Therefore the treatments targeted towards reduction of the oral malodour will require antimicrobial components directed against the tongue microbiota.

Treatment of OM is important not only because it helps patients to achieve self-confidence but also because the evidence indicates that VSC's can be toxic to periodontal tissues even when present at extremely low concentrations [35]. The best way to treat OM is to ensure that patients practice good oral hygiene and that their dentition is properly maintained [36] . Traditional procedures of scaling and root planing can be effective for patients with OM caused by periodontitis[37] . All patients should be instructed in proper tooth brushing and flossing and tongue cleaning. Mouthrinses should be recommended based on scientifically proven efficacy. Caution should be exercised and professional advice should be sought as to administration and type of mouthrinse to be used. Tongue scraping should be demonstrated and patients should be asked to demonstrate to the dental hygienist the appropriate use of tongue scrapers. The tongue has a tendency to curl up while tongue scraping therefore a combination of flexible tongue scrapers and tongue scrapers with handles should be recommended to the patients.
The saliva functions as an antibacterial, antiviral, antifungal, buffering and cleaning agent [38] and so any treatment that increases saliva flow and tongue action, including the chewing of fibrous vegetables and sugarless gum, will help to decrease OM [39]. Finally, oral rinses can be used as supplement to good oral hygiene practices.

Antimicrobial agents

Mouthwashes have been used as chemical approach to combat oral malodour. Mouth rinsing is a common oral hygiene measure dating back to ancient times [39]. Antibacterial components such as cetylpyridinium chloride, chlorhexidine, triclosan essential oils, quaternary ammonium compounds, benzalkonium chloride hydrogen peroxide, sodium bicarbonate [41], Zinc salts and combinations of these compounds have been considered along with mechanical approaches to reduce OM [40]. Any successful mouthrinse formulation must balance the elimination of the responsible microbes while maintaining the normal flora and preventing an overgrowth of opportunistic pathogens. Most commercially available mouthrinses only mask odours and provide little antiseptic function. Even when these mouthrinses do contain antiseptic substances, the effects are usually not long-lasting [42][43] The microbes survive antiseptic attacks by being protected under thick layers of plaque and mucus [13].

Many commercially available rinses contain alcohol as an antiseptic and a flavor enhancer. The most prevalent problem with ethanol is that it can dry the oral tissues. This condition in itself is a risk factor for OM. In addition, there is some controversy as to whether or not the use of alcohol rinses are associated with oral cancer [44][45] .The FDA states that there is no evidence to support the removal of alcohol from over-the-counter products but alcohol-free mouthrinses are becoming increasingly popular.

Zinc Rinses: Clinical trials conclude that zinc mouthrinses are effective for reducing OM in-patients with good oral health [39] Zinc rinses (in chloride, citrate or acetate form) have been found to reduce oral VSC concentrations for longer than three hours. The zinc ion may counteract the toxicity of the VSC's and it functions as an odour inhibitor by preventing disulfide group reduction to thiols and by reaction with the thiols groups in VSCs. It has been reported that a zinc-based rinse was more effective as compared to chlorine dioxide based rinse when both rinses were used twice a day for 60 seconds over a 6-week period [46] . Zinc containing chewing gum has been shown to reduce oral malodour [47]. Chlorhexidine Rinses: Chlorhexidine digluconate is useful in decreasing gingivitis and plaque formation. It is one of two active ingredients in mouthrinses that has been shown to reduce gingivitis in long-term clinical trials and appears to be the most effective anti-plaque and anti-gingivitis agent known today [13][48].

The efficacy of chlorhexidine as a mouthrinse to control OM has not been studied extensively. The primary side effect of chlorhexidine is the discoloration of the teeth and tongue. In addition, an important consideration for long-term use is its potential to disrupt the oral microbial balance, causing some resistant strains to flourish, such as *Streptococcus viridans* [49]

The effect of 1-stage full mouth disinfection in periodontitis patients (Scaling and root planing of all pockets within 24 hrs together with the application of chlorhexidine to all intra-oral niches followed by chlorhexidine rinsing for 2 months) resulted in a significant improvement in oral malodour when compared to a conventional periodontal therapy [50]. While chlorhexidine appears to be clinically effective from these open-design clinical studies, it is not an agent that should be used routinely, or for long periods of time, in the control of oral malodour, because of its side effects. Mouth rinse containing chlorhexidine, cetylpyridium chloride and zinc lactate was evaluated in a clinical study for two weeks. Eight subjects participated in this pilot study and this formulation showed improvement in organoleptic scores and a trend to reduce tongue and saliva microflora [51] . Antimalodour properties of chlorhexidine spray 0.2% chlorhexidine mouthrinse 0.2% and sanguinaria – zinc mouthrinse were evaluated on morning breath. Oral malodour parameters were assessed before breakfast and four hours later after lunch. Results indicated that a sanguarine –zinc solution had a short effect as compared to chlorhexidine that lasted longer [52] .Chlorine Dioxide Rinse : Chlorine dioxide (ClO_2) is a strong oxidizing agent that has a high redox capacity with compounds containing sulfur. Chlorine dioxide is also used in water disinfection and in food processing equipment sanitation and functions best at a neutral pH. Commercially available mouthrinses are a solution of sodium chlorite since chlorine dioxide readily loses its activity [39]. Further independent clinical investigations are needed to substantiate the effectiveness of sodium chlorite containing rinses for the control of OM. In fact, chlorine dioxide the agent most widely touted on the Internet has no published clinical studies to substantiate the claims to reduce oral malodour. Benzalkonium chloride in conjunction with sodium chlorate has been shown to be effective in reducing oral malodour.

In this pilot study subjects with mild to severe periodontitis were instructed to use the mouthwash twice a day for a period of six weeks and periodontal and oral malodour parameters were assessed [53] Results indicated a trend towards reduction in pocket depths and bleeding on probing.

Triclosan Rinses: Triclosan (2, 4, 4'-trichloro-2'-hydroxydiphenylether) is a broad spectrum nonionic antimicrobial agent. This lipid soluble substance has been found to be effective against most types of oral bacteria [54]. A combined zinc and triclosan mouthrinse system has been shown to have a cumulative effect, with the reduction of malodour increasing with the duration of the product use [39]. Two-Phase Rinses: Two-phase oil-water mouthrinses have been tested for their ability to control OM. A clinical trial reported significant long-term reductions in OM from the whole mouth and the tongue dorsum posterior [55]. The rinse is thought to reduce odour-producing microbes on the tongue because there is a polar attraction between the oil droplets and the bacterial cells. The two-phase rinse has been shown to significantly decrease the level of VSCs eight to ten hours after use, although not as effectively as a 0.2% chlorhexidine rinse [56]. Positive controls such as Chlorhexidine and Listerine that had previously shown to reduce organoleptic scores were used in these clinical studies. Hydrogen Peroxide : The potential of hydrogen peroxide to reduce levels of salivary thiol precursors of oral malodour has been investigated. Using analytical procedures reduction in salivary thiols levels post treatment was found to be 59 %. [57].

Topical antimicrobial agent+ Azulene ointment with a small dose of clindamycin was used topically in eight patients with maxillary cancer to inhibit oral malodour that originates from a gauze tamponade applied to the postoperative maxillary bone defect. The malodour was markedly decreased or eliminated in all cases. Anaerobic bacteria such as *Porphyromonas* and *Peptostreptococcus* involved in generation of malodour also became undetectable [58] .

Other products

Breathnol is a propriety mixture of edible flavors, which was evaluated, in a clinical study and this formulation reduced oral malodour for atleast 3 hours [59] . Lozenges, Chewing gums, Mints: Oxidizing Lozenges have been reported to reduce tongue dorsum malodour [60] Alternative remedies:Some of the natural controls for oral malodour include gum containing tea extract. Also recommended are natural deodorants such as copper chlorophyll and sodium chlorophyllin. Alternative dental health services also suggest the use of chlorophyll oral rinses in addition to spirulina and algae products to contain antimicrobial properties.

Conclusions

Many of the mouthrinses available today are being used for the prevention and or treatment of oral malodour. Much more research is required to develop an efficacious mouthrinse for the alleviation of oral malodour. The treatment of oral malodour is relatively a new field in dentistry and many of the treatments have involved a trial and error approach, but the knowledge and experience gained so far will hopefully facilitate clinical investigations in this field and eventually lead to improved diagnostic techniques and treatment products.

References:
[1]   Nachnani S, Oral Malodour: A Brief Review CDHA Journal 1999; 14 (2): 13-15
[2]   Survey conducted at ADA reveals interesting trends. Dent Econ 1995; Dec: 6.
[3]   Gerlach R W, Hyde J.D., Poore C.L. et al Breath effects of three marketed Dentrifices: A comparative study evaluating single use and cumulative use. The journal of clinical dentistry , Volume 1X , Number 4, 1998.
[4]   Scully C, Porter R, Greenman J. What to do about halitosis? Brit Med J 1994; 308: 217-218.
[5]   Preti G; Clark L; Cowart BJ et al. Non oral etiologies of oral malodour and altered chemosensation. J Periodontol 1992 Sep; 63 (9): 790-96.
[6]   Young K, Oxtoby A. Field EA. Halitosis: A review. Dental Update 1993,20 57-61.
[7]   Rosenberg M. Clinical assessment of bad breath: current concepts J Am Dent Assoc 1996 Apr; 127(4): 475-82
[8]   Touyz LZ. Oral Malodour – a review. J Can Dent Assoc 1993 Jul; (7): 607-610
[9]   Ship JA. Gastatory and olfactory considerations. Examination and treatment in   general practice. J Am Dent Assoc 1993; 31:53-73.
[10]   Tiomny E, Arber N Moshkowitz M, Peled Y, Gilat T. Halitosis and *Helicobacter pylori*. J Clin Gastroenterl 1993; 15: 236-237.
[11]   Klienberg I, Westbay G. Salivary metabolic factors and involved in oral malodour formation. J periodont 1992 Sep; 63 (9): 768-75
[12]   Newman M.G. The Role of periodontitis in Oral Malodour: Clinical Perspectives Bad breath: A multidisciplinary approach. van Daniel Steenberghe & Mel Rosenberg Editors.
[13]   Rosenberg M, Editor. Bad Breath: research perspectives. Ramot Pub. Tel Aviv, 1995.
[14]   Tessier JF, Kulkarni GV. Bad Breath: etiology and treatment. Periodontics Oral Health. Oct   1991; 19-24.
[15]   Tonzetich J. Production and origin of oral malodour: a review of mechanisms and methods of analysis. J Periodontol, 1977; 48 (1): 13-20.
[16]   Yaegaki K,   Sanada K. Biochemical and clinical factors influencing oral malodour in periodontal patients. J   Periodontol, 1977; 48 (1) 13-20.
[17]   Clark G, Nachnani S, Messadi D, CDA Journal Vol 25, No 2,February 1997
[18]   Goldberg S, Kozlovsky A, Gordon D, Gelernter I, Sintov A, Rosenberg M. Cadaverine as a putative component of oral malodour. J Dent Res 1994 Jun; 73(6): 1168-72.
[19]   Waler S. On the transformation of sulfur-containing amino acids and peptides to volatile sulfur compounds (VSC) in the human mouth. Eur J Oral Sci 1997,105 (5 pt 2): 534-537
[20]   Yaegaki K, Sanada K.: Volatile sulfur compounds in mouth air from clinically healthy subjects and patients with periodontal disease. J Periodont Res 1992; 27:223-238
[21]   De Boever EH, Loesche WJ. The tongue microflora and tongue surface characteristics contribute to oral malodour In: van Steenberge D. Rosenberg M (eds), Bad Breath: A Multidisciplinary Approach. Leuven Belgium: Leuven University Press, 1996:111-121.
[22]   Kozlovsky A; Gordon D; Gelernter I; Loesche WJ; Rosenberg M Correlation between BANA test and oral malodour parameters. J Dent Res 1994 May; 73(5): 1036-1042
[23]   De Boever EH, Loesche WJ, Assessing the contribution oaf anaerobic microflora of the tongue to oral malodour. JADA 1995; 126:1384-1393.
[24]   Klienberg L, Codipilly M, and the biological basis of oral malodour formation In Rosenberg M. (Ed) Bad Breath: research Perspective. Tel Aviv, Israel Ramot Publishing. Tel Aviv University: 1995:13-39.

[25]   McNamara TF, Alexander JF and Lee M. The role of microorganisms in the production of oral malodour. Oral Surg 1972; 34-41.

[26]   Goldberg S, Kozlovsky A, Gordon D, Gelernter I, Sintov A, Rosenberg M. Cadaverine as a putative component of oral malodour. J Dent Res 1994 Jun; 73(6): 1168-1172

[27]   Rosenberg M, Septon I, Eli I, Bar-Ness A, Gelenter, I Bremer, S Gabbay J. Halitosis measurement by an industrial sulfide monitor. J Periodon1991;62: 487-489.

[28]   Rosenberg M, Kulkarni GV, Bosy A, McCulloch CAG: Reproducibility and sensitivity of oral malodour measurements with a portable sulfide monitor. J Dent Res 1991; 11:1436-1440.

[29]   Doty RL, Green PA, Ram C, Yankel SL. Communication of gender from human breath odours: Relationship to perceived intensity and pleasantness. Norm Behav 1982.16:13-22

[30]   Tonzetich J. Richter VJ. Evaluation of volatile odiferous components of saliva. Arch Oral Biol 1964; 9: 39-45

[31]   Gibson.TD, Prosser O, Hulbert JN. Et al. Detection and simultaneous identification of microorganisms from headspace samples using an electronic nose. Sensors and Atuators 1970: B 44:413-22

[32]   Neiders M, Brigette Ramos. Operation of bad breath clinics. Quintessence International Proceedings of the third International Conference on Breath Odour 1999; 30(5): 295-301.

[33]   Deems DA, Doty RL, Settle RG, Moore-Gillon V, et al. Smell and taste disorders, a study of 750 patients form University of Pennsylvania Smell and taste Centre. Arch Otolaryngeal Head and Neck Surg 1991; 117: 519-528

[34]   Delanghe G, Ghyselen J, Feenstra L, Van Steenberghe D. Experiences in of a Belgium multidisciplinary breath odour clinic Bad Breath – A multidisciplinary Approach. Leuven, Belgium Leuven University Press 1996:199-208.

[35]   Johnson PW, Yaegaki K, Tonzetich J. Effect of volatile thiol compounds on protein metabolism by human gingival fibroblast. J Periodont Res 1992; 27: 533-561.

[36]   Rosenberg M. Bad Breath: Diagnosis and treatment. Univ Tor Dent. J. 1990; 3(2): 7-11.

[37]   Ratcliff PA, Johnson JW. The relationship between oral malodour, gingivitis, and Periodontitis. A review. J Periodont Res 1999; 70 (5): 485-489.

[38]   Spielman Al, Bivona P, Rifkin BR. Halitosis: A common oral problem. N.Y. Sate Dent J 1996 Dec; 63 (10): 36-42.

[39]   Nachnani S, The effects of oral rinses on halitosis. CDA Journal,1997;25-28.

[40]   Mandel I.D. Chemotherapeutic agents for controlling plaque and gingivitis. J. Clin Periodontol 1988; 488.

[41]   Grigor J, Roberts AJ. Reduction in the levels of oral malodour precursors by hydrogen peroxide in vitro and in vivo assessments. J. Clin Dent 1992; 3(4):111-115.

[42]   Moneib NA, El- Said MA, Shibi AM. Correlation between the in vivo and in vitro antimicrobial properties of commercially available mouthwash preparations. J Chemotherapy, 4(50): 276-280, 1992.

[43]   Pitts G, Brogdon C, Hu L, Masurat T, Pianotti R, Schumann P. Mechanism of action of an antiseptic, anti-odour mouthwash. J Dent Res, 1983; 62 (60): 738-742.

[44]   Gagari E, Kabani S. Adverse effects of a mouthwash use. Oral Surg. Oral Med Oral Pathol Radiol Endod.1995; 80(4): 432-439.

[45]   Elmore JG, Horwitz RJ. Oral cancer and mouthwash use: Evaluation of the epidemiological evidence. Otolaryngology- Head and Neck Surgery, 113(30 253-261,

[46]   Nachnani S, Anson D. Effect of Orasan on periodontitis and oral malodour. (Abstract), J Dent Res 1998; 77(6).

[47]   Nachnani S. Reduction of oral malodour with zinc containing chewing gum (abstract). J Dent Res1999; 78.

[48]   Beiswanger BB, Mallet ME, Jackson RD et al Clinical effects of a 0.12% chlorhexidine rinse as an adjunct to scaling and root planing. J Clin Dent, 3 (2): 33-38,1992.

[49]   Lauri H, Vaahtoniemi LH, Karlqvist K, Altonen M, Raisanen S. Mouth rinsing with chlorhexidine causes a delayed temporary increase in the levels of oral viridans streptococci. Acta Odontol Scand, 1995; 53:226-229.

[50]   Quirynen M; Mongardini C; van Steenberghe D. The effect of 1-stage full mouth disinfection on oral malodour and microbial colonization of the tongue in periodontitis. A pilot study. J Periodontol 1998 Mar; 69(3): 374-382

[51]   Roldan S, Herrera D, & Sanz M Clinical and Microbiological Effects of an antimicrobial Mouthrinse in Oral Mouthrinse. Abstract Fourth International Conference on Breath odour UCLA 1999.

[52]   D van Steenberghe, Avontroodt B, and Vandekerkhove. A comparative Evaluation of a chlorhexidine Spray, a Chlorhexidine Mouthrinse and a Sanguinarine-Zinc Mouthrinse on Morning Breath Odour. (abstract) Fourth International conference on Malodour.

[53]   Nachnani S, Anson D. Effect of Orasan on periodontitis and oral malodour. (Abstract), J Dent Res 77(6) 1998.

[54]   Gaffar A. Sheri D, Afflitoj, Coleman EJ. The effect of triclosan on mediators of gingival inflammation. J Clin Periodontol, 22 (60 480-484, 1995.

[55]   Kozlovsky A, Goldberg S, Natour I, Rogatky-Gat A, Gelernter I, Rosenberg M. Efficacy of a two phase oil water mouthrinse in controlling oral malodour, gingivitis and plaque. J Periodontol, 1996; 67(6): 577-578.

[56]   Rosenberg M, Gelentre I, Barki M, Barness. Daylong reduction of oral malodour by two-phase oil water mouthrinse as compared to chlorhexidine and placebo rinses. J Periodontol, 1992; 63 (1); 39-43.

[57]   Grigor J; Roberts AJ. Reduction in the levels of oral malodour precursors by hydrogen peroxide in vitro and in vivo assessments. J. Clin Dent 1992; 3(4): 111-115.

[58]   Ogura T; Urade M, Matsuya T. Prevention of malodour from intraoral gauze with the topical use of Clindamycin. Oral Surg Oral Med Oral Pathol 1992 Jul; 74 910: 58-62.

[59]   Rosenberg M, Barkim, Goldberg –Levitan et al. Oral reduction By Breathanol. (abstract) Fourth International conference on Breath Research. UCLA Los Angeles. 1999.

[60]   Greenstein RB, Goldberg S, Marku-Cohen S, Stere N, Rosenberg M. Reduction of oral malodour by oxidizing lozenges. J Periodontal 1997 Dec; 68 (12): 1176-1181

Taste Quantification Using the Electronic Tongue

A. Legin, A. Rudnitskaya, B. Seleznev, Yu. Vlasov

Chemistry Department, St.Petersburg University, Universitetskaya nab. 7/9,

199034 St.Petersburg, Russia

Introduction

The analysis of liquid media, containing various organic substances, namely different kinds of beverages and foodstuffs is an urgent analytical task. In the recent years a promising approach to these problems based on the utilization of sensor arrays and pattern recognition/multivariate analysis methods ("electronic tongue") has been suggested [1,2]. This device combines the advantages of single sensors with new interesting abilities. Thus the electronic tongue is capable to perform multicomponent quantitative analysis and classification or recognition of complex media [3 and reference therein]. One of the most exciting features of the electronic tongue is the possibility to establish correlations between its output and human perception of taste and/or flavour. To provide meaningful and reliable information about complicated multi component systems such as foodstuffs, the sensors of the array must exhibit reproducible sensitivity to organic substances. The present paper is devoted to evaluation of sensor sensitivity to organic substances typical for food and comparison of electronic tongue output with human perception.

Experimental

Sensitivity of different sensing materials to typical organic substances from different classes, usually present in foods, has been tested. The substances were alcohols, organic acids, phenols, ethers, aldehydes, terpenes etc. Also sensitivity to some typical "basic taste" substances, such as quinine, urea, and glutamate has been evaluated. The concentration ranges were from 10^{-5} to 10^{-1} mol/L and even higher for selected substances. Calibrations in the solutions of each substance have been repeated at least 5 times, typical reproducibility of the sensor response were 1-3 mV. Two main types of sensing materials have been studied: solid-state chalcogenide glasses and PVC based organic polymer membranes, totally over 100 different compositions. All sensors have been produced in the Laboratory of Chemical Sensors of St.Petersburg University, sensor membranes compositions are given elsewhere [4,5].

Measurements with the electronic tongue including 21 sensors have been made in individual solutions and binary mixtures of 3 taste substances used for taster panel "calibration": Taste substances were sodium chloride, lactic acid and L-leucine, which have salty, sour and bitter taste respectively. Simple solutions contained only one substance at 3 concentration levels, binary mixtures contained 2 of the taste substances at different concentration levels. Sensors displaying sensitivity to organic compounds have been included in sensor arrays for measurements in foodstuffs.

The electronic tongue including 30 sensors has been used for measurements in coffee brews and different sodas. Measurements using the "electronic tongue" have been made in 8 individual sorts of coffee beans. Ground coffee samples have been obtained from the manufacturer along with taste quality grading chart made by professional taste panel. The brew has been prepared according the standard procedure using an automatic drip coffeemaker and cooled down to the room temperature before measurements. According to the recommendation of the manufacturer the brew concentration were 23.3g of ground coffee per 1000 ml of the distilled water.

The measurements using the "electronic tongue" have been made in 8 samples of soft drinks: 4 samples of regular and diet Pepsi and Coke and 4 experimental ones of Pepsi. All samples have been obtained from the US manufacturer together with organoleptic estimates of "dietness" degree, made by the tasters of the company. All sodas have been stirred before measurements to eliminate the excess of CO_2.

Different multivariate analysis and pattern recognition methods have been used for experimental data processing. Recognition of simple and binary solutions of taste substances has been done using principal component analysis. Taste intensity estimation in taste substances binary mixtures has been done using partial least square regression, measurements in simple solutions being used for calibrations. Correlation between the "electronic tongue" data and taster panel estimation of coffee and soft drinks has been performed by back-propagation neural network using live-one-out methods.

Results and discussion

It was found that many sensors, developed in our lab display reproducible sensitivity to organic compounds that can be present in different foodstuffs. Sensitivity of chalcogenide glass and PVC based sensors to different alcohols and phenols are shown in Tables 1 and 2. Sensitivity of PVC sensors to taste substances such as sodium glutamate (umami) and quinine (bitter) are shown in Fig. 1. It is important to point out that sensors display sensitivity to organic ion itself, e.g. response slope to glutamate is negative that corresponds to anion sensitivity, response slope to quinine is positive. Sensors with sensitivity to organic substances have been incorporated into the array of electronic tongue, which was successfully used for quantitative analysis and qualitative recognition of different foodstuffs.

An attempt to make taste quantification using the electronic tongue has been performed. Results of single taste substances discrimination using the electronic tongue are shown in the Fig. 2. Three groups of samples are very clearly distinguished. Inside each group the samples with 3 different concentrations of taste substances are well separated. Results of binary mixtures discrimination are shown in the Fig. 3. Three groups of binary mixtures can be clearly distinguished. Afterwards taste quantification has been performed. The electronic tongue has been calibrated using the data acquired in single taste substance solutions and measurements in binary mixtures have been used as tests. The electronic tongue has correctly predicted which substances and at what concentration level have been present in each binary mixture (Table 3).

The electronic tongue has been applied for quantitative beverage flavour evaluation. Professional taste panel scores have been used as reference for electronic tongue. One half of the samples have been used as calibration ones, another half as test samples. Taste panel evaluated coffee using 4 parameters: flavour, smell, body and acidity, each of them ranking from 1 to 10. Results of coffee estimates prediction by the electronic tongue in test samples are shown in the Table 4. Taster panel evaluated soda samples according to one parameter – "dietness", regular sodas being anchored to 10, diet ones to 2 or 3. Taste difference between regular and diet sodas is related mostly to different sweeteners used, natural or artificial non-carbohydrate. Results of sensory estimates prediction by the electronic tongue in soda test samples are shown in the Table 5. Thus, the electronic tongue is capable to correctly predict sensory scores of flavour features in beverages, i.e. the electronic tongue output can be correlated to human flavour (taste) perception.

Conclusions.

Sensitivity of chalcogenide glass and PVC plasticized sensors of different compositions to organic and taste substances have been studied. The electronic tongue was capable to distinguish different taste substances in individual and binary solutions and estimate taste intensity (concentration) of each substance in the mixture. Correlation between the "electronic tongue" output and human perception (grading chart produced by professional taste panel) has been observed for coffee and soft drinks.

Acknowledgments. Authors wish to thank Dr. Patrick Mielle and Dr. Cristian Salles (INRA, Dijon, France) for supply of taste substance samples and fruitful discussions.

References.

[1] Di Natale C, D'Amico A, Vlasov Yu G, Legin A V, Rudnitskaya A M In Proc.of the Intern.

Conf. EUROSENSORS IX Stockholm Sweden 1995 512 (news).

[2] Hayashi K, Toko K, Yamanaka M, Yoshihara H, Yamafuji K, Ikezaki H, Toykubo R, Satj K

1995 Sensors and Actuators B **23** 55-61.

[3] Vlasov Yu, Legin A 1998 Fresenius J. Anal. Chem. **361** 255-260.

[4] Vlasov Yu G, Bychkov E A, Legin A V 1994 Talanta **41** No6 1059-1063.

[5] Legin A, Rudnitskaya A, Smirnova A, Lvova L, Vlasov Yu 1999 J. of Applied Chemistry

(Russian) **72** Issue1 114-120.

Table 1. Sensitivity of sensors to different alcohols.

Sensors	Sensor rersponse, mV/decade					
	ethanol	ethyleneglycol	glycerine	isoamyl	iso-propanol	propanol
ch1	17.5	-27.4	-5.4	-20.4	-10.3	-9.2
ch2	16.8	-21.9	-7.5	-11.4	-13.2	-9.5
ch3	8.8	-20.5	-5.3	-19.1	-8.6	-7.6
pvc1	-7.2	-9.8	0	-7.6	0.6	-0.3
pvc2	-7.9	-6.5	0.4	-4.7	2.8	-0.2
pvc3	16	4.5	0.7	6.4	1.1	1.2
pvc4	-19.8	-14	-1.1	-10.8	-2.5	-3
redox	32.1	-17.7	1.6	-15.9	-4.9	-3.3
pH	1.5	-1.5	0.1	-1.2	0.9	0.4

Table 2. Sensitivity of sensors to different phenols.

Sensors	Sensor sensitivity, mV/decade			
	phenol	4-ethylphenol	eugenol	2-metoxiphenol
pvc1	68.9	61.9	15.8	35.8
pvc2	61.7	50.2	13.2	28.1
pvc3	41.5	55.2	10.4	15.6
pvc4	32.5	43.2	8.2	12
pvc5	30.7	47	8.5	12.7
pvc6	30	47.9	2.5	14.8
pvc7	23.5	38.5	4	6.5
pvc8	19	30	2	8.5
pvc9	16	31.7	5	2.5
pvc10	18	29	6.2	3.5
pvc11	17	24.7	4.5	5.7
pvc12	16.5	32.3	3.2	1.5
redox	-3.5	11.2	14.7	8
pH	-4.5	-2.4	9.9	-4.4

Table 3. Taste quantification using the electronic tongue.

Sample	Sourness		Bitterness		Saltiness	
	real	found	real	found	real	found
B1	low	low	low	low	0	0
B2	low	medium	0	0	low	medium
B3	0	0	low	0	low	medium
B4	high	high	high	high	0	0
B5	high	high	0	0	high	high
B6	0	0	high	low	high	high
B7	high	high	0	0	low	medium
B8	0	0	high	medium	low	low
B9	0	0	low	low	high	high

Table 4. Coffee parameters evaluation.

Factor	Sample	5	6	7	8
Flavour	El.tongue	4.8±0.2	7.8±0.1	3.1±0.3	4.1±0.4
	Tasters	3-7	7-9	1-2	2-4
Acidity	El.tongue	4.8±0.1	3.6±0.3	5.7±0.3	5.4±0.3
	Tasters	5-8	2-5	5-8	5-8
Body	El.tongue	7.0±0.1	8.5±0.01	8.4±0.1	6.0±0.1
	Tasters	5-9	8-9	8-9	5-7
Smell	El.tongue	5.3±0.2	8.4±0.01	3.8±0.3	4.7±0.3
	Tasters	4-7	8-9	2-3	3-5

Table 5. Evaluation of soda "dietness" using the electronic tongue.

Sample	Tasters	El. tongue	St.deviation
350	3	3.0	0.2
830	4.5	4.6	0.2
852	7	6.9	1.1
410	3	6.2	0.4
Pepsi diet	2	2.3	0.2
Pepsi regular	10	9.0	0.2
Coke diet	3	2.2	0.3
Coke regular	10	9.6	0.1

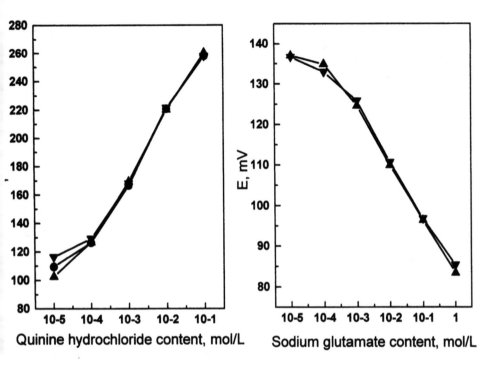

Fig. 1. Sensitivity of sensors to taste substances.

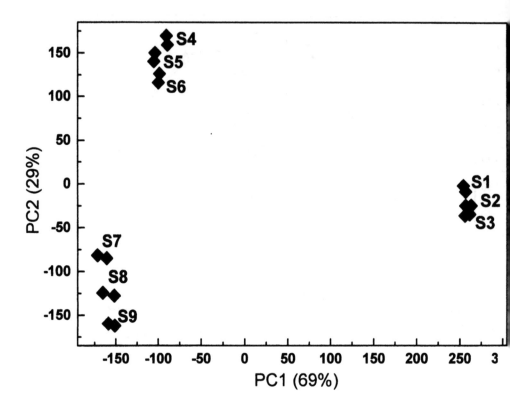

Fig. 2. Discrimination of taste substances; S1, S2, S3 - sodium chloride solutions (salty taste), S4, S5, S6 - L-leucine solutions (bitter taste), S7, S8, S9 - lactic acid solutions (acid taste).

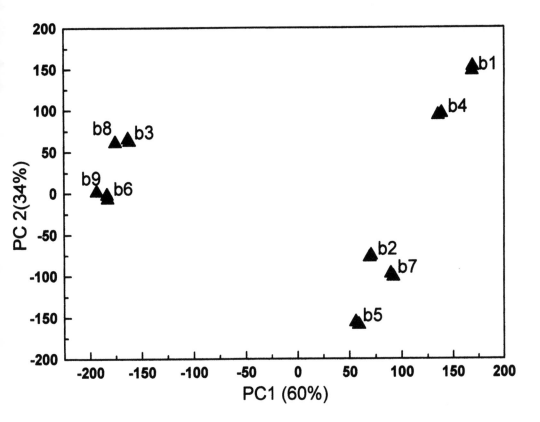

Fig. 3. Discrimination of taste substance binary mixtures; B1, B4 – mixtures of lactic acid and L-leucine (sour and bitter taste), B2, B5, B7 – mixtures of sodium chloride and lactic acid (sour and salty taste), B3, B6, B8, B9 – mixtures of sodium chloride and L-leucine (bitter and salty taste).

Section 1: Odours, Taste and Physico-chemical Interactions
Paper presented at the Seventh International Symposium on Olfaction and Electronic Noses, July 2000

23

A Boiling Point Model to Describe Conducting Organic Polymer Gas Sensor Responses

Richard A. Bissell*[1], Feng-Bin Li[2], Paul Travers[1] and Krishna Persaud[2]

1. Osmetech plc, Electra Way , Electra House, Crewe CW1 1WZ, UK.
2. DIAS, UMIST, PO Box 88, Manchester, M60 1QD, UK.
E-mail: mcnvrab@fs3.in.umist.ac.uk

Abstract. Through assessment of the steady-state electrical resistance changes of poly(3,3″-didecyl-2,2″:5′,2″-terthiophene) doped with various counter ions upon exposure to multiple concentrations of 27 volatile organic chemicals, it has been shown that a linear relationship exists between $\log(1/C_v)$ and T_b (C_v being the analyte concentration required to generate a fixed sensor response, and T_b being analyte boiling point) for analytes which share a common functional group. The study supports a physico-chemical model of conducting organic polymer gas sensor responses based on non-specific partition of the analyte between vapour and polymer phases.

1. Introduction

Conducting organic polymers (COPs), typically polypyrrole (PP), have been widely used as sensing elements for the detection of volatile organic chemicals (VOCs) [1]. Whilst the mechanism of sensor action [2] has received some attention, little work has been done to correlate COP sensor responses with analyte physico-chemical parameters. Such correlations would enable the prediction of COP sensor responses to untested analytes as well as providing a more complete description of COP sensor performance for studies of sensor mechanism. In the case of surface acoustic wave (SAW) gas sensor devices it has been shown that sensor responses can be correlated to analyte boiling point [3]. Establishing the correlation was facilitated by the ease with which partition coefficients can be obtained from SAW sensor responses. In the case with COP sensors, however, no similar results have been reported to date, perhaps due to the different physical properties measured and the complex transduction processes [4-6] involved.

In this paper we report linear relationships between steady-state COP sensor responses and analyte boiling point for groups of analytes which share the same functional group. The COP materials investigated are of the polyalkylthiophene (PAT) type. Very few reports on the use of doped PATs in gas sensing have been published [8]. PATs are attractive materials for applications in molecular electronics [9] because of their stability in both the doped and undoped states, and good processability.

2. Theoretical background and the boiling point model

2.1. Relationship between gas partition coefficient and analyte boiling point

The theory of molecular partition between the gas phase and a condensed phase is well established in the field of gas liquid chromatography (GLC) [10]. The partition coefficient, K, which is defined as

$$K = C_c/C_v \tag{1}$$

(where C_c is the concentration of the vapour in the chromatographic stationary phase and C_v is the concentration of the vapour in the carrier gas phase) depends on the solubility of the vapour in the polymer which in turn is a function of the saturated vapour pressure of the vapour and the strength of the vapour-polymer interactions [3]. In GLC studies the partition coefficient has been shown to follow the well-known expression [10]

$$K = RT(\rho_1/M_1)/(\gamma_2 p_2) \qquad (2)$$

Where ρ_1 and M_1 are the density and molecular weight of the stationary phase, respectively, p_2 is the saturation vapour pressure of the solute vapour and γ_2 is the coefficient of interaction between vapour and polymer ($\gamma_2 = 1$ for ideal solutions).

The following approximate expression, based on Trouton's rule and the Clausius-Clapeyron equation relates p_2 and boiling point T_b

$$\log p_2 \approx 7.7 - T_b (t/2.303RT) \qquad (3)$$

where t is the Trouton constant for the vapour.

Combining equations 2 and 3 yields

$$\log K = a + bT_b \qquad (4)$$

The parameters a and b in Equation 4 are defined as $a = \log (RT/\gamma_2) -7.7 + \log (\rho_1/M_1)$ and $b = t/(2.303RT)$. For an ideal vapour-polymer solution at a given temperature, constant a is a function of the polymer alone (through the term ρ_1/M_1, $\gamma_2 \approx 1$), and constant b of analyte alone (through the parameter t, which is in the range 85-90 J·mol^{-1}·K^{-1} for non-hydrogen bonding vapours [9]). Deviations from ideality arise when $\gamma_2 \neq 1$ due to specific vapour-polymer interactions.

2.2. *PAT sensor responses and VOC partition coefficients*

It can be assumed that the sensor response is proportional to the concentration of the analyte gas dissolved in the polymer:

$$\Delta R = \%dR/R = C_c \cdot c_t \qquad (5)$$

where ΔR is the percentage change in resistance of the PAT sensor upon exposure to the VOC, C_c is the concentration of VOC dissolved in the polymer and c_t is a 'transduction constant' indicating the sensitivity with which the amount of the partitioned analyte is translated by the polymer film into a resistance change.

By combining equations 1 and 5 the following equation is obtained

$$K = (\Delta R /c_t) /C_v \qquad (6)$$

Since the transduction mechanism is complex, it is not possible at present to directly calculate c_t. However, studies by Kunugi *et al* [11] of simultaneous mass and resistance changes in PP films upon exposure to a homologous series of alcohols (methanol-butanol) have shown that i) for each alcohol the relative mass change was proportional to the relative resistance change, which is consistent with equation 5, and ii) that the relative resistance change per adsorbed molecule on the PP film was essentially constant across the series of alcohols investigated. This implies that the *transduction constant, c_t, is the same across this series*. If the gas concentrations that induce a fixed level of response in the sensor are obtained then equation 6 reduces to

$$K \propto 1/C_v \qquad (7)$$

Where C_v is the gas concentration required to produce a fixed level of response in the sensor for a series of compounds, which share the same functional group, e.g. alcohols.

Finally by combining equations 4 and 7 the following expression is obtained

$$\log (1/C_v) = A + a + bT_b \qquad (8)$$

where A is an arbitrary constant which depends on the response value chosen in equation 6 and is equal to log $(\Delta R \cdot c_t)$.

3. Experimental

The sensors investigated comprise a pair of thin (~1μm) gold electrodes vapour deposited on a ceramic substrate with a separation of 250 μm. A *p*-xylene solution (8 mg/mL) of poly(3,3''-didecyl-2,2'':5',2''-terthiophene) - subsequently referred to as DEC - was drip cast (50 μL over an area of ~ 2 cm^2) over the electrodes and subsequently doped electrochemically with 0.1M solutions of Bu_4NPF_6, $Bu_4NCF_3SO_3$ and Bu_4NClO_4 in acetonitrile. The sensor materials are referred to as DEC/dopant ion. DEC was characterised by ^{1}H nmr spectroscopy and gel permeation chromatography (M_w = 22,900, M_n = 8,050). The thickness of the films are 1-3 μm. Resistance changes at 35.0 (±0.05) °C in air (relative humidity 50%) on exposure to sample vapours were monitored using an Aromascan analyser (A32S) and proprietary software. Analyte gas samples were generated at 22.0 (±0.1) °C. Accurate gas concentrations (± 2%) were generated by diluting saturated vapours using a custom built automated gas blending rig. All transfer tubes were teflon. The percentage steady-state change in sensor resistance (%dR/R) was measured at four gas concentrations with three repeats at each concentration. The data thus obtained were fitted using a polynomial function (R^2 values typically > 0.9). The gas concentration required to give a response of 0.25 %dR/R was calculated from the calibration curve. In all cases this response value lay in the linear range of sensor responses and was at least three times greater than the minimum reliably detected signal (ca. 0.05% dR/R). All data are averages of three sensors. A total of 27 analytes were investigated comprising four chemical classes: *alcohols* (methanol, ethanol, propanol, pentanol, hexanol, octanol, propan-2-ol, 3-methyl-1-butanol, butan-2-ol, cyclohexanol, 2-methylpentan-2-ol, 2-ethyl-1-hexanol), *esters* (ethyl hexanoate, butyl acetate, ethyl acetate, ethyl butyrate, ethyl octanoate, isoamyl acetate), *aromatic hydrocarbons* (mesitylene, *o*-xylene, *p*-xylene, toluene, ethyl benzene, butylbenzene), *aliphatic hydrocarbons* (octane, decane, dodecane). All analytes were at least 98% pure and were used as received.

4. Results and discussion

4.1.   Features of the sensor response

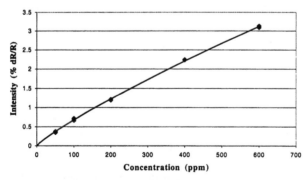

Figure 1. Percentage change in steady-state resistance of DEC/ClO$_4$ in response to various concentrations of propanol (three repeats at each concentration).

All the sensor responses to all the analytes reported in this paper were reversible and positive, i.e. resistance increased upon exposure to the vapour and returned to a steady baseline value after exposure was ended. A typical relationship between steady-state sensor response and gas concentration (in this case propanol vapour) is shown in Figure 1. The calibration plot is almost linear over the entire concentration range tested, which is typical of all the sensor/analyte profiles.

4.2. Response to alcohols and the boiling point model

Figure 2 shows a plot of $\log(1/C_v)$ *vs.* T_b (see equation 8) for sensor DEC/ClO$_4$ in response to a variety of alcohols including primary, secondary, tertiary and cyclic species spanning a wide range of boiling points. It is evident that a linear correlation is observed. The correlation is strongest if only homologous primary alcohols are considered ($R^2 = 0.99$, n=6) but the inclusion of non-homologous species does not greatly reduce the correlation ($R^2=0.97$, n=13).

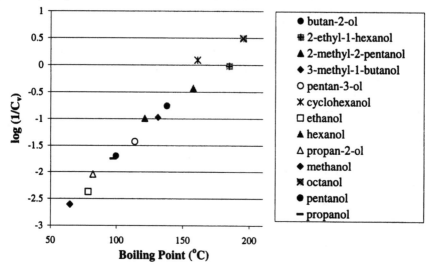

Figure 2. Log $(1/C_v)$ *vs.* T_b for sensor DEC/ClO$_4$ in response to a variety of alcohols.

Thus it is apparent that the sensor does not discriminate between alcohols of widely differing size and shape. An exception is perhaps the cyclic alcohol, cyclohexanol, for which for all the sensors studied gave larger signals than expected. This observation may be related to the lower entropies of vaporisation of cyclic compounds relative to their open-chain counterparts [9].

Figure 3 displays the responses of DEC doped with a variety of counter ions to the same series of alcohols. It is clear that the linear log $(1/C_v)$ – T_b correlation in figure 2 is not limited to one particular dopant ion. The linear relationships in Figure 3 have similar gradients but show more variation in the intercepts. The former observation is predicted by the boiling point model which states that ideally the gradient is not dependent on the polymer type but is related to the Trouton constant (entropy of vaporisation) of the analyte (see Section 2). Calculation of Trouton's constant from the gradients in Figure 3 provides reasonable values in the range 96-121 J· mol^{-1}· K^{-1} (literature values for alcohols range between 102 and 112 J· mol^{-1}· K^{-1} [9]).

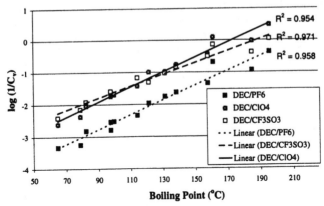

Figure 3. Log $(1/C_v)$ *vs.* T_b for variously doped DEC in response to a variety of alcohols.

Since the gas concentrations in Figure 3 are the values required to generate a fixed sensor response (0.25% dR/R), the lower the required vapour concentration the more sensitive is the sensor. The variation in the intercepts in Figure 3, therefore, indicates that the sensors have different sensitivities to alcohols. The fact that the dopant ion affects the sensitivity of COP sensors is well known [12], although it is clear that in the present example improved sensitivity does not result in increased sensor selectivity for different alcohols.

It is noteworthy that equations similar to 8 have long been observed in medical studies where correlations for related drugs exist between experimentally observed quantities such as the minimum lethal dose (LD_{50}) and empirical parameters associated with the drugs such as the octanol-water partition coefficient (Hansch parameter) [13]. In both cases, the mechanism of interaction of the drug/organism or VOC/gas sensor is complex, but overall the process is governed by non-specific physico-chemical partition, which can be described by simple free energy relationships.

4.3. *Responses to other analyte classes*

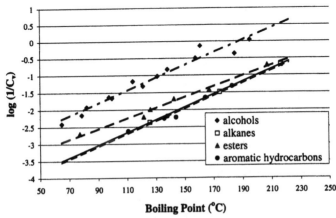

Figure 4. Log $(1/C_v)$ *vs.* T_b for sensor DEC/ClO$_4$ in response to four functional groups.

Figure 4 shows that the linear log $(1/C_v)$ – T_b correlation is observed for other analyte types as well as alcohols. Further, in the studied boiling point range, the sensitivity order of sensor DEC/ClO$_4$ towards the analyte classes investigated is alcohols > esters > aromatic hydrocarbons ≈ aliphatic hydrocarbons. Similar observations were made for DEC sensors doped with hexafluorophosphate and triflate. The sensitivity order is consistent with the expected relative partition coefficients of the four analyte classes with a polar material since polar molecules are expected to have larger free energies of interaction than non-polar molecules. In the present model this is reflected in a non-ideal coefficient of mixing ($\gamma_2 < 1$) of polar vapours with the polymer (equation 4). The gradients of the relationships in Figure 4 are very similar to each other which reflects the fact that the entropies of vaporisation of most non-strongly hydrogen bonding liquids are roughly constant (Trouton's rule).

5. Summary

A model of COP sensor performance based on the partition of VOCs between vapour and polymer phases has been described. The model has been tested for a polyalkylthiophene material (DEC) doped with various counter ions. The steady-state changes in electrical resistance of the material upon exposure to multiple concentrations of 27 vapours covering four different chemical groups has provided support for the model. The general conclusions of the study are that i) DEC is more sensitive to polar rather than nonpolar analyte types, ii) DEC does not display marked size or shape selectivity towards a group of analytes which share the same functional group but differ widely in secondary structure, e.g. homologous and non-homologous alcohols, iii) changing the dopant ion in DEC has a large effect on *absolute* sensor sensitively but does not markedly affect the *relative* sensitivity towards different analyte functional groups (alcohols > esters > hydrocarbons). The model and the general conclusions are considered to applicable to other types of COPs.

References

[1]   Hatfield J V, Neaves PI, Hicks P J, Persaud K C and Travers P
        1994 Sens. Actuators B - Chem. 18 221-8
[2]   Slater M S, Paynter J and Watt E J 1993 Analyst 118 379-84
[3]   Patrash J P and Zellers E T 1993 Anal. Chem. 65 2055-66
[4]   Blackwood D and Josowicz M 1991 J. Phys. Chem. 95 493-502
[5]   Slater M S, Watt E J, Freeman N J, May I P and Weir D J 1992 Analyst 117 1265-70
[6]   Gardener J W, Bartlett P N and Pratt K F E
        1995 IEE Proc.-Circuits and Devices Syst. 142 321-33
[7]   DeWit M, Vanneste E, Blockhuys F, Geise HJ, Mertens R and Nagels P
        1997 Synth. Met. 85 1303-4
[8]   Bao Z, Rogers J A and Katz H E 1999 J. Mater. Chem. 9 1895-1904
[9]   Schwarzenbach R P, Gschwend P M and Imboden D R
        1993 Environmental Organic Chemistry (Wiley-Interscience) Chapter 4
[10]   Littlewood A B 1970 Gas Chromatography (New York:Academic) 44-121
[11]   Kunugi Y, Nigorikawa K, Harima Y and Yamashita K
        1994 J. Chem. Soc. Chem. Commun. 873-4
[12]   Topart P and Josowicz M 1992 J. Phys. Chem. 96 7824-30
[13]   Meyer H 1899 Arch. Exp. Pathol. Pharmakol. 42 109-18

Section 2

Sensors/Instrumentation

Section 2: Sensors/Instrumentation
Paper presented at the Seventh International Symposium on Olfaction and Electronic Noses, July 2000

31

Low Frequency A.C. Polymeric Gas Sensor

G. C. do Nascimento[a], R. Souto-Maior[b], C. P. de Melo[b]

[a]ITEP – Instituto Tecnológico do Estado de Pernambuco, Recife, Brazil
[b]Universidade Federal de Pernambuco, Recife, Brazil

Abstract. We show that the low frequency region of the impedance spectra of poly-3-thiophene acetate (P3ETA) films is very sensitive to the presence of volatile compounds. This fact can be used for the development of electronic noses, which contrary to the usual practice of polymeric gas sensors, would use a single P3ETA sample instead of an array of differently modified conducting polymer films.

1 Introduction

It is well known that the exposition to vapors of volatile compounds affect in a remarkable manner the electrical properties of films of conducting polymers. Although the complete details of the microscopic interaction between the adsorbed molecules and the polymeric sample are still not completely understood, the specificity of the D.C. response is good enough to allow the development of electronic noses, instruments able to identify the nature of the volatile species after comparing the pattern of changes in the electrical characteristics of an array of similar - but differently doped - conducting polymer films exposed to the odorant molecules. In this work, we call attention to the fact that the low frequency region of the impedance spectra of poly-3-thiophene acetate films exhibits not only a high sensitivity but also a very fast response to the presence of volatile compounds. We suggest that the wealth of information that can be extracted from its A.C. characteristics make this polymer a very promising material for the development of an electronic nose based on a single film rather than on multiple polymeric samples.

2 Experimental

2.1 Synthesis of the material

The poly-3-ethylthiopheneacetate (P3ETA) films were prepared by electrochemical synthesis in a closed cell under argon atmosphere [1]. The ethyl-3-thiopheneacetate monomer was freshly distilled previous to use and nitrobenzene and lithium perchlorate of analytical quality (Aldrich, USA) were used as electrolytes. The polymer was formed on the surface of indium tin oxide (ITO) electrodes by applying a constant current of 0.2 mA/cm^2 for 45 minutes, and the good mechanical properties of the self standing films made easy to peel them off the substrates. Small samples (4mm x 5 mm) cut from these films were used for the four-point electrical measurements.

2.2 Electrical Characterization

By definition, in an alternate current circuit the impedance $Z(\omega)$ of a device for a given frequency ω is the relation

$$Z(\omega) = V(\omega) / I(\omega)$$

between the electrical voltage $V(\omega)$ applied and the resulting current $I(\omega)$. The impedance of the polymeric samples was obtained by applying an external excitatory electrical step pulse (with an intensity of 2 Volts) and mathematically analyzing the frequency spectrum of the response. As an event with a short duration in time, the external pulse corresponds to a very broad band of excitatory frequencies. However, the polymeric sample will only absorb energy at certain specific frequencies, and therefore given its A.C. response will necessarily be more well defined. Due to the large value of the sample's resistance ($\sim$ 6MΩ), the resulting current is very low and a linear Ohmic behavior can be assumed for the polymeric sample. For this linear analysis of frequencies [2], a Fourier transform algorithm was used to convert the temporal response in the corresponding frequency spectrum. The entire acquisition process was done in an automatic manner, and the electronic circuit [3] was entirely enclosed in a metallic box to avoid electrical interferences.

3 Results

Vapors of acetone, ethyl-acetate, methanol, ethanol and water were used for the tests. For each individual analysis, 10 mL of the chosen substance were allowed to evaporate in a Petri dish inside the metallic box, and 16 transient response were collected, each consisting a set of data sampled at 20 sec intervals after the substance had been injected a once. In Fig. 1 we show the results for ethanol at room temperature, and the curves that corresponds to the responses collected right before the first, and after the 16-th transient, respectively.

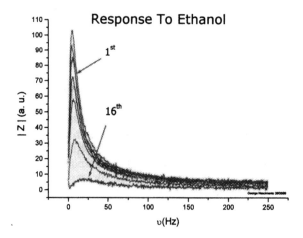

Fig1. Low frequency spectra of a P3ETA film, after exposition to ethanol.

In a similar manner, the impedance spectrum of the P3ETA films rapidly changes in a very noticeably way after the exposition to the vapors of the other compounds. However, each of these responses has its own temporal evolution as can been seen in Fig. 2.

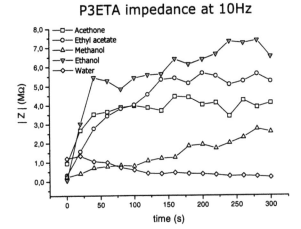

P3ETA impedance at 10Hz

Fig2. *Temporal P3ETA impedance response for various volatile at 10Hz.*

As a consequence, after a preliminary step of calibration, where specific answers would be associated to individual substances, an aroma sensor based on a single type of polymeric films can be developed.

4 Conclusions

We show that just one sample for the P3ETA films exhibit a very fast and specific A.C. response to the exposition of different vapors. There are no noticeable signal response at higher frequency and this suggest the possibility that these phenomena are related with viscoelastic relaxation at low frequency, as it is well known for others class of polymers [4]. Different regions for the same P3ETA has similar relaxation response and this is a good indication about the homogeneity for the synthesized film. The broad range of the observed changes in the impedance spectrum of these films can be exploited for the development of electronic noses capable of operating as real time instruments.

5 References

[1] Souto-Maior R 1988 *Phd Dissertation – University of California, Santa Barbara.*
[2] Lathi P B 1968 Signals, Systems, and Communication. (New York: Jonh Willey & Sons, Inc.)
[3] Fafilek G and Breiter M W 1997 *Journal of Eletrochemical Chemistry* **430** 269-78
[4] McCrun N G, Read B E and Willians G 1967 *Anelastic and Dielectric Effects in Polymeric Solids.* (New York: Dover Publications Inc)

Characterisation of an Electrodeposited Conducting Polymer FET Array for Vapour and Odour Sensing

J A Covington[1], J W Gardner[1], C Toh[2], P N Bartlett[2], D Briand[3] and N F de Rooij[3]

[1]University of Warwick, School of Engineering, Coventry, CV4 7AL, UK
[2]University of Southampton, Dept. of Chemistry, Southampton, SO17 1BJ, UK
[3]University of Neuchatel, Institute of Microtechnology, Neuchatel, CH-2007, Switzerland

Abstract. Electrochemically-deposited conducting polymers have been used in electronic noses to detect volatile organic compounds for a number of years. Here we investigate the use of poly(pyrrole) and poly(bithiophene) as the gate material of a ChemFET, and so produce an electrodeposited multi-polymer FET array. Results show that these ChemFET devices are sensitive to ethanol and toluene vapour with a sensitivity coefficient of typically 0.8 µV/PPM and 8 PPM limit of detection. We believe that the response is associated with a modulation of the workfunction between the polymer and the gate oxide.

1. Introduction

The integration of standard CMOS technology with conducting polymer resistive sensors has been of interest to researchers for many years. Such a combination should lead to a low-cost, low power gas sensor for the handheld electronic nose market. Recent research into conducting polymer resistive [1] and FETs (PolFETs) [2,3] has demonstrated their sensitivity to volatile organic compounds, although they tend to suffer from significant humidity dependence and long-term drift. Here we investigate further electrodeposited conducing polymers (CPs) as the gate material of a MOSFET, its response to two analytes and the effects of ambient humidity and temperature. We also report upon a ChemFET array device employing conducting polymers, onto which different electrodeposited conducting polymers have been deposited. Using this technique, and with the wide range of conducting polymers available, it is possible to create arrays of CMOS compatible PolFET devices suitable for handheld electronic noses.

2. FET devices and polymer deposition

The silicon die employed here comprises of an array of 4, enhanced FET sensors, with a single *p-n* thermodiode, fabricated at the Institute of Microtechnology, University of Neuchatel, Switzerland. The FET sensor has either a solid gold gate electrode, covering the channel area (closed) or etched away exposing the gate oxide (open), allowing the polymer to be in direct contact with the insulating oxide layer. The FET sensors have a combined gate and drain with typical channel dimensions of 10 μm $\times$ 300 μm or 5 μm $\times$ 150 μm (channel length/channel width) and an 80 nm thick gate oxide. The channel of the FETs is meandered to improve electrical connectivity to the polymer film and to reduce the size of the sensing area. These NMOS devices are fabricated in a *p*-well, with *n*-type substrate to inhibit any cross-interference between devices [4]. A schematic of a sensor is shown in figure 1a. The devices are passivated with either an oxynitride or nitride PECVD film. The sensors are configured in two ways, initially with two open and two closed gates, i.e. where only one polymer material was being used (duplicated) or with four open gates for the multi-polymer sensor array.

The conducting polymers used for these experiments were poly(pyrrole)/butane sulfonic acid, sodium salt (BSA) and poly(bithiophene)/tetrabutylammonium tetrafluoroborate (TFBTFA), deposited at the Department of Chemistry, University of Southampton, UK.   The gold gates of the FET were cleaned using cyclic voltammetries in 1 mol dm^{-3} sulphuric acid. This was carried out by employing a micro-deposition technique in which a luggin capillary and counter wire were combined into a single structure [5]. This technique was modified by adding an adjoining arm to the luggin capillary from which the level of solution, hence the height of solution above the device, can be adjusted. Thus the pressure exerted by the solution above the device was controlled. This modification was necessary when the solution used was acetonitrile which has a lower surface tension than water, and tends to flow out of the luggin capillary. The luggin capillary was filled with 1 mol dm^{-3} sulphuric acid and connected to a double-frit SCE reference electrode. A column of electrolytic solution, maintained by its surface tension, from the luggin capillary was then carefully positioned over the centre sensing area of the PolyFET device using the *xyz* translator of the microstage.

Poly(pyrrole) was deposited from an aqueous solution containing 0.3 mol dm^{-3} pyrrole and 0.10 mol dm^{-3} BSA. Electrodeposition of poly(pyrrole) was carried out using potential step from 0 V to 0.85 V (vs. SCE). The amount of polymer deposited was controlled by the amount of charge passed. Poly(bithiophene) was deposited from an acetonitrile solution containing 0.04 mol dm^{-3} 2,2'-bithiophene and 0.10 mol dm^{-3} TBATFB. A double-frit Ag/AgNO$_3$ reference electrode replaced the double-frit SCE as the reference electrode. Electrodeposition of poly(bithiophene) was carried out using potential steps from 0 V to 0.85 V for 1 s, followed by potential step to 0.815 V (vs. Ag/AgNO$_3$).

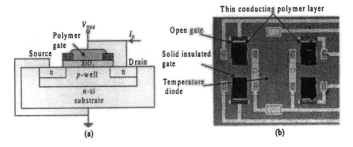

(a)   (b)

Figure 1: Schematic (a) and photograph (b) of the PolFET sensor structure.

The thickness of the polymer films was profiled using a Wyko™ NT-2000 interferometer. This instrument uses optical phase shifting and vertical scanning interferometry to create a three-dimensional surface profile of the polymer. Typical thickness ranged from 2 μm to 10 μm depending on the deposition time. A photograph of the sensing area after polymer deposition is shown in figure 1b. This shows the definition of the deposition process where only the gate electrodes and channels are covered in polymer.

3. Experimental set-up

The sensors were placed in an automated test station consisting of custom made electronics, mass flow controllers (MFCs), valves and a PC running National Instruments™, LabVIEW version 5.0. Purified air is passed through a fine sinter within the analyte, saturating this carrier gas with the analyte. This is diluted by further purified air, in which a fixed amount of water vapour has been added, measured by a Rotronic HP-103A humidity sensor. This combination at a fixed concentration of sample vapour and humidity is passed over the sensors. The concentration and humidity level is controlled by the flow rate through the sample vessel and at the mixing stage, with a maximum flow rate of 300 ml/min. Further details are published elsewhere [6].

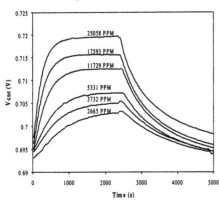

Figure 2: Typical responses of PolFET sensor to ethanol.

The sensors where placed in a DRI-BLOCK™ heater for a period of 24 hours to thermally equilibrate before testing and at a humidity level of typically 30% r.h. The devices were driven at a constant current of 10 μA and the drain/gate voltage (V_{GDS}) is monitored with respect to the source. The sensors were exposed alternatively to pulses of ethanol and toluene vapour of between 1000-20000 PPM, for ethanol and 1000-8000 PPM of toluene in air for 30 min followed by a 40 minute recovery period and at four humidity levels (3000 to 9800 PPM of water, equivalent of 10-50% r.h. at 20 °C).Also the effect of temperature on the sensors was investigated with tests performed at temperatures between 25 to 60 °C. Typical responses of poly(bithiophene) to ethanol is shown in figure 2 at a fixed temperature of 25 °C and humidity level of 4500 PPM of water. Three repeats of each sensor was tested, with a typical set of responses shown in the results section.

4. Results

Initial tests performed showed that both types of material responded to these analytes. Typical sensitivities to ethanol and toluene were measured at +0.8 to +1.5 µV/PPM and −0.2 to −0.7 µV/PPM for poly(bithiophene)/TBATFB and +0.5 to +0.7 µV/PPM and −0.05 to −0.1 µV/PPM for poly(pyrrole)/BSA, respectively. The responses following a Langmuir isotherm, as shown in figure 3. The observed time response of the sensors was considerable (> 5 min) but that the response was seen to be reaction-rate not diffusion-rate or test rig limited.

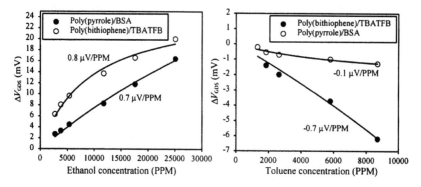

Figure 3: Response of PolFET sensor fitted to Langmuir isotherm for ethanol and Toluene at a fixed temperature of 30 °C.

Also investigated was the effect of humidity on both the baseline signal and on the magnitude of the response. If we consider first the effect of humidity on the response, then a reduction in V_{GDS} was observed for an increase in humidity at fixed temperature, as shown in figure 4a. The results of these tests shown a Langmuir dependence to humidity with typical values of −0.9 to −1.7 µV/PPM for poly(pyrrole)/BSA and −3.1 to −3.6 µV/PPM for poly(bithiophene)/TBATFB.

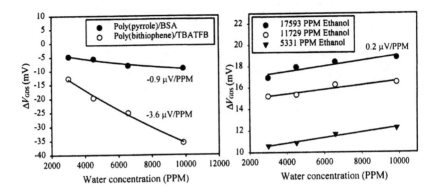

Figure 4: (a) Effect of water vapour on baseline, (b) effect of water vapour on response at a fixed temperature of 30 °C.

The effect of humidity on the magnitude of the response was also examined, whereupon it was found, curiously, that an increasing humidity level led to an increase in response. This increase was found to be typically linear with an increase in response measured at typically +0.2 μV/PPM of water vapour at constant vapour concentration.

The last measured effect was that of temperature on the magnitude of the response. It was observed that increasing temperature caused a reduction in the response, which exhibited a linear dependence over the test range, as shown in figure 5.

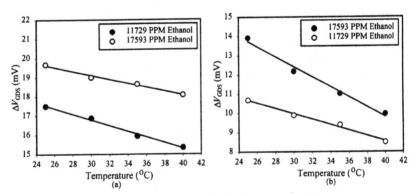

Figure 5: Effect of temperature of magnitude of response for (a) poly(bithiophene)/TBATFB and (b) poly(pyrrole)/BSA.

Table 1 summarises the coefficients of concentration, humidity and temperature on these PolFET devices, including the sample variance.

Polymer	$\left.\dfrac{dV_{GDS}}{dC_E}\right\|_{T=25°C}$ in µV/PPM	$\left.\dfrac{dV_{GDS}}{dC_T}\right\|_{T=25°C}$ in µV/PPM	$\left.\dfrac{dV_{GDS}}{dC_H}\right\|_{T=25°C}$ in µV/PPM	$\left.\dfrac{dV_{GDS}}{dT}\right\|_{C_E}$ in µV/°C
Poly(pyrrole)/BSA	+0.6 ±0.2	-0.1 ±0.1	-1.3 ±0.5	125 ± 35
Poly(bithiophene)/TBATFB	+1.4 ±0.6	-0.4 ±0.2	-3.4 ±0.4	201 ± 54

Table 1: Typical coefficients observed for the PolFET sensors

5. Analysis and discussion

If we first consider a standard MOSFET working in the saturated region then the forward *I-V* characteristic can be modelled by:

$$i_{DS} = g_m \left(V_{GDS} - V_T\right)^2 = \kappa \tag{1}$$

where g_m is a conductance-related term (not the transconductance), i_{DS} is the drain/source current, κ is a constant and V_T is the threshold voltage.
By re-arranging this equation, and assuming constant current and temperature then we can relate V_T to V_{GDS} by:

$$\Delta V_T = \Delta V_{GDS} \tag{2}$$

If we consider this further then the response of these sensors can be equated to a shift in the flatband potential of the FET device, specifically a shift in the workfunction between the polymer and the silicon (ϕ_{PS}). Here we believe that exposure of analyte to the sensor causes a modification in the workfunction within the chemFET device, which is supported by other research [7]. A number of different possible mechanisms could account for this response. Firstly it could be due to the analyte behaving as a donor (or acceptor) of electrons to (from) the polymer; however it is unlikely that charge transfer takes place here with organic analytes. Secondly, this change in work function could be associated with a bulk swelling effect of the polymer, i.e. as the polymer swells the volume fraction of the polymer falls. This could modify the surface density of states of the polymer system and hence the effective workfunction. Lastly the analyte may alter the height of the potential barrier between polymeric molecules (rather than distance) and thus modify the local band structure of the polymer – this occurs throughout the film and so scales up for a macroscopic effect. Here the potential height of bands is altered modifying the interaction between polymer chain and the counter ion. This is a localised effect, where these affects over the polymer combine to modify the workfunction.
        Results have shown a Langmuir isotherm type of response to vapour concentration, which suggests that there is a fixed number of independent absorption sites to which the ethanol molecules can bind. The shift in threshold voltage can be expressed as:

$$\Delta V_T = \frac{a_E K_E C_E}{1 + K_E C_E} \tag{3}$$

where K_E is a binding coefficient, C_E is the ethanol concentration and a_E is a constant. A similar response is observed for toluene, but shows that the sign of the response is analyte dependent.

An effect of humidity on both the baseline and response was observed. The baseline also follows a Langmuir isotherm, as given by in eqn. 3, but with water vapour instead of analyte. A possible explanation for these observed results is due to the formation of a dipole layer between the water molecules and the dangling oxygen bonds on the gate oxide. This dipole moment will add to the previously applied gate voltage shifting the characteristics of the PolFET sensor.

The experimental results suggest that poly(bithiophene)/TBATFB is more porous to water molecules causing an increased shift. The effect of humidity on the magnitude of the response was also investigated. Here it was observed that increasing the humidity level produces a larger response. This is unlike similar experiments with resistive based conducting polymer sensors where the water and ethanol are competing for the same binding site causing a reduction in the response of the sensor. At present we have no definitive explanation for this behaviour.

As described earlier an increase in temperature causes a linear reduction in response. This decrease is significantly lower that would be expected from a resistive conducting polymer sensor, where the fall in response follows an exponential function. A possible explanation for this reduction is given by Hang et al. [8], where an increase in temperature reduces the solubility of the analyte in the bulk of the sensors material. This mechanism is a simple thermodynamic law, basically described as a reduction of the analyte concentration in the polymer as the temperature increases. Thus, defining a coefficient K_{PG} as the ratio of the analyte concentration in the polymer ($c_{i,poly}$) to the analyte concentration in the gas ($c_{i,gas}$), gives:

$$K_{PG} = \frac{c_{i,poly}}{c_{i,gas}} \tag{4}$$

This may be shown from basic thermodynamics to be related to ambient temperature by:

$$\log(K_{PG}) \propto \frac{4.4T_b}{T} \tag{5}$$

Where T_b is the boiling point of the solvent. This assumes an ideal mixing of molecules and entropy of evaporation of 85 $JK^{-1}mol^{-1}$. We can now re-write the response of the sensor to ethanol vapour as:

$$\Delta V_T \propto \frac{a_E K_{PG} K_E C_E}{1 + K_{PG} K_E C_E} \approx \frac{a_E \exp\left(\dfrac{4.4T_b}{T}\right) K_E C_E}{1 + a_E \exp\left(\dfrac{4.4T_b}{T}\right) K_E C_E} \tag{6}$$

at constant humidity. Over the temperature range used for these experiments this can be approximated to a linear function, as observed experimentally. This is obviously a simplified solution because the interaction between the polymer and the analyte will also be temperature dependent but is assumed here to be a negligible.

6. Conclusion

We have described a novel FET sensor array for the detection of odours and vapours. The sensors have employed two conducting polymers, poly(pyrrole)/BSA and poly(bithiophene)/TBATBF as the gate material of these FET devices. Sensitivity to ethanol and toluene vapour in air was observed to follow a Langmuir isotherm of up to +0.8 μV/PPM – this suggests a limit of detection of around 8 PPM of vapour. Humidity dependence of the baseline (zero gas) signal was observed, which also followed a Langmuir isotherm; possibly due to the formation of a dipole layer between the water vapour and the gate oxide. Furthermore the unexpected effect of increased response with increased humidity was observed. Lastly the temperature dependence of the sensors was measured and accounted for by a change in the solubility of the analyte in the bulk polymer.

The number of conducting polymers available and the ease of deposition at room temperature make these materials attractive for chemical sensors. Integration of sensor structures with a standard CMOS process may lead to low cost, reproducible sensors for the handheld electronic nose market.

References
[1]   Covington J A, Gardner J W and Hatfield J V 1999 SPIE 6th Annual International Symposium on Smart Structures and Materials 296-307
[2]   Ingleby P, Gardner J W, Bartlett P N 1999 Sensors and Actuators B 57 17-27
[3]   Meijerink M G H, Koudelka M, de Rooji N F, Strike D J, Hendrikse J, Olthuis W and Bergveld P 1999 Electrochemical and Solid-state letters, 2(3) 138-139
[4]   Briand D, van der Schoot B, Sundgren H, Lundström I and de Rooij N F 1999 Tech. Digest of Transducers'99 Sendai, Japan, 938-941. To be published in J. Microelectromech. Syst., vol 9(3), 2000
[5]   Bartlett P N, Elliott J M and Gardner J W 1997 Measurement and Control, Vol. 30 273-279
[6]   Covington J A, Gardner J W and Hatfield J V 1999 Proc. of SPIE 3673 296-307
[7]   Hang M, Schierbaum K D, Gauglitz G and Gopel W, Sensors and Actuators B, 11 (1993) 383-391

Acknowledgements
The author would like to acknowledge the EPSRC for their funding of this research.

Section 2: Sensors/Instrumentation
Paper presented at the Seventh International Symposium on Olfaction and Electronic Noses, July 2000

43

Grading of Coating Materials from 2D-τ-charts of Gas Sensor Transient Response to Microencapsulated Flavour

A. Šetkus[1], R. Bocevičiutė[2], A. Galdikas[1], Ž. Kancleris[1], D. Senulienė[1], P. R. Venskutonis[2]

[1] *Semiconductor Physics Institute, A.Goštauto 11, Vilnius, Lithuania
Phone: +370-2-627934, Fax: +370-2-627123, E-mail: setkus@uj.pfi.lt*

[2] *Department of Food Technology, Kaunas University of Technology, Radvilėnu pl. 19, Kaunas, Lithuania*

Abstract. Transient response of the resistance is investigated in metal oxide based gas sensors exposed to microencapsulated flavours. The 2D-τ-chart is composed of parameters extracted from the time dependent signal by exponential approximation. Coating materials used for the encapsulation and the essential oils encapsulated are characterised basing on the 2D-τ-charts and gas chromatography.

1 Introduction

In an electronic nose, stationary signals usually are collected from number of sensors exposed to an odour. The target gas could be recognised if special processing is applied to the collection. It was proved recently [1] that the transient response of the sensors is much more efficient for the recognition of smell features than the stationary signal. In a new approach, a time dependence of the signal is analysed in the sensors exposed to a steep change in gas composition of air. Improved analysis of the transient response is applied for studying of flavours in present report.

Traditional dried ground spices and herbs, although being widely used products possess several serious disadvantages such as variable flavour strength and profile, contamination with filth, etc. Microencapsulation has become an attractive method to solve the problems. During the encapsulation, thin films or polymer coats are applied to small solid particles, droplets of liquids or gases [2]. Characteristics of the encapsulated flavour are highly dependent on the coating material used.

Present report deals with an evaluation of the coating materials based on an analysis of transient response to gas in metal oxide gas sensors and on standard chromatography.

2 Experimental tests and analysis

2.1 Microencapsulation

The essential oils from dried plant materials were prepared by traditional hydrodistillation method in a Clevenger type apparatus. The following essential oils from the dried herbs, seeds and commercial powders (cassia) were prepared for the tests: 1) thyme (*Thymus vulgaris* L.); 2) cassia (*Cinnamomum cassia* Presl); 3) oregano (*Origanum vulgare* L. spp. *hirtum*). Selected essential oils were encapsulated in different coating materials (matrixes) by preparing emulsions and subsequent spray drying in a laboratory instrument. For the coating we used maltodextrins from National Starch & Chemical, USA: Hi Cap 100 produced from waxy maize, Capsul E produced from tapioca, Encapsul 855 produced from waxy maize, N LOK produced from waxy maize. The volatile compounds released from encapsulated oils were analysed by capillary gas chromatography (GC) and coupled GC and mass spectrometry (GC/MS). Solid phase microextraction (SPME) was employed in the analysis.

2.2 Gas sensor tests

An array of gas sensors was exposed to the volatile constituents released from the encapsulated oils. In the two-flow control system, switching between the flow of clean air and that of the target smell produced a steep change in gas composition of air. A time dependence of the resistance change was scanned for three home-made sensors (SO18-23, SO18-24, ITO16) based on tin and indium oxides and one standard TGS-800 sensor from the Figaro. The scanning rate was fixed at 1 point per second in all tests.

2.3 Analysis of the transients

An integration method originally proposed by Tittelbach-Helmrich [3] was used for the analysis of the measured transients. We implemented a program for the transient's analysis based on the integration method. Two least-square problems are solved using SVD procedure, the roots of polynomial are found using the Muller's method [4]. In the practical procedure of the signal analysis we performed the calculations by increasing the number of components extracted from the transient until the complex roots of the polynomial is found. To have an idea of the accuracy of the method we performed some statistical tests using artificially generated transients with Gaussian noise added. Time constants and weighting coefficients were plotted for each object in the 2D-τ-charts.

3 Results

3.1 Transient response of gas sensors

After an injection of the odour from the encapsulated oils, the resistance change was scanned with similar rate for all the sensors. Typical transient responses are illustrated for one of the sensors in figure 1. In the figure, the resistance (R_{oil}) decreased in response to the oils at the beginning and tended to a stabile value after that. Long lasting changes were detected in some transients (e.g. cassia in figure 1). Though the clean air resistance (R_{air}) was recovered by restoring the clean air flow, an injection series was limited by a poisoning of the gas system after long flow of the odours.

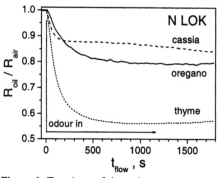

Figure 1. Transients of the resistance responses of the SO18-23 sensor to odours emited from the essential oils encapsulated in the N LOK matirx.

The transients similar to that in figure 1 fitted with the approximation

$$R_{oil}/R_{air} = \sum_{i=0}^{M} a_i \exp(-t/\tau_i). \qquad (1)$$

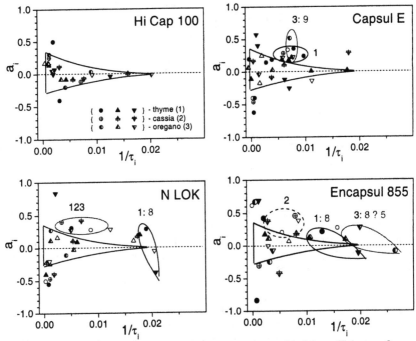

Figure 2. The 2D-τ-charts composed of time constants τ_i and weigh coefficients a_i from transient responses of three gas sensors (SO18-23, SO18-24 and TGS800) to odour released from the essential oils (thyme, cassia and oregano) encapsulated in the matrixes: Hi Cap 100, N LOK, Encapsul 855 and Capsul E.

Based on this approximation, each transient was characterised with few pairs (M) of related parameters namely the time constant τ_i and weigh coefficient a_i. Depending on the matrix and oil, there were extracted from two to four pairs of the parameters. Each pair (τ_i, a_i) defined a point in an individual chart of a coating. In figure 2, the 2D-τ-charts are illustrated for each of the matrixes probed. In each of the charts in figure 2, the results are summarised for all used essential oils and the sensors. Reciprocal time scale was used to plot τ_i because short and intermediate time constants were the target in most cases. Consequently, the long τ-components were concentrated near the zero and are purely distinguished in the charts in figure 2. A vertical distance from the zero-line upward or downward in the chart illustrates the weigh coefficients.

Based on the smallness of the weigh coefficients, the Hi Cap 100 was supposed to be the least leaking matrix in the tests. For this matrix the (τ_i, a_i)-points were spread within the smallest area around the zero-line in the 2D-τ-chart. The outlines of this area were like an arrow directed along the τ-axis (figure 2, Hi Cap 100). This smallest area was also drawn in other charts in figure 2. It is evident that only part of the (τ_i, a_i)-points were inside of this smallest area in the 2D-τ-charts of other matrixes. The outside clusters of the points revealed the differences between the matrixes. For the Capsul E the outside clusters were obtained from the transient responses to thyme and oregano. These two clusters (1 and 3) nearly covered each other on the chart (figure 2, Capsul E).

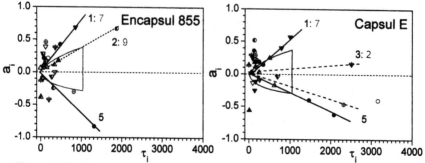

Figure 3. The 2D-τ-charts (long time constant view) based on the transient responses of the sensors to the odours of the oils encapsulated in the Encapsul 855 and Capsul E.

Two outside clusters were also found in the chart of the N LOK. The cluster 1 of short time constants corresponded to thyme encapsulated in the N LOK. The weigh coefficients of the short components were big enough to make this cluster clearly seen in the chart. The intermediate time cluster marked by 123 in figure 2 consisted of the points corresponding to all the oils tested in our work. According to the 2D-τ-charts, the Encapsul 855 appeared to be the most leaking matrix in our study (figure 2). In this chart, the outside clusters were located individually for each oil encapsulated in the Encapsul 855. The shortest τ-constants were extracted from the transient response to oregano (3 in the Encapsul 855 chart in figure 2). Short τ-constants were also obtained for thyme (cluster 1) though these constants were longer than for oregano (cluster 3 in the Encapsul 855 chart in figure 2). The cluster 2 of the (τ_i, a_i)-points was found for cassia in the interval of intermediate time constants. Some weigh coefficients in this cluster were exceeding the reference area twice or even more in the a_i-scale.

From some transient responses, very long τ-constants were extracted for the matrixes N LOK, Encapsul 855 and Capsul E. These long τ-constants are illustrated in the charts with normal scaling for the τ_i-axis in figure 3. The (τ_i, a_i)-points obtained for the same sensor were scattered near the beam-like lines in the charts. It should be noted, that each beam-like line was related to the τ-component detected for different oil. Only the long τ-components labelled by 1 were present in the approximation of all three matrixes.

3.2 The SPME analysis

The concentrations of the main compounds above the microencapsulated oils were obtained using the SPME. Few different absorbents were used in these tests. In the table 1, however, the absolute amounts are tabulated for the main compounds absorbed by non-polar polydimethylsiloxane.

According to the results in table 1, the Hi Cap 100 was the least leaking matrix. An amount of the volatile compounds was negligible for the tested oils above this matrix. Emission of the volatile compounds from other three matrixes was much greater than from the Hi Cap 100. According to the main compounds, the composition of the odour emitted from cassia were evidently different from the composition of other two oils encapsulated in any of the matrixes. Thyme and oregano were very close to each other by the composition of the main compounds in table 1. One from the four compounds was different for each of these two oils. The concentrations of the emitted compounds were also comparable for thyme and oregano in the leaking matrixes.

Table 1. Concentration of main volatile compounds above microencapsulated essential oils from the SPME tests. Non-polar polydimethylsiloxane absorbent was used.

		Hi Cap 100 a.u.·10^6	Capsul E a.u.·10^6	Encapsul 855 a.u.·10^6	N LOK a.u.·10^6
Cassia	Benzaldehyde	0.13	1.42	3.21	0.53
	(E)-Cinnamaldehyde	nd	14.86	nd	10.27
	Cinnamyl acetate	0.297	3.28	3.52	2.23
	p-Methoxycinamaldehyde	0.067	1.7	1.60	1.41
Thyme	p-Cymen	-	10.42	26.60	6.7
	Borneo	-	2.48	3.75	4.15
	Thymol	0.15	18.22	18.25	nd
	Carvacrol	0.04	7.28	13.8	12.65
Oregano	p-Cymene	0.06	5.78	20.60	1.09
	Limonene	0.067	2.44	0.34	0.64
	Thymol	0.223	3.08	9.20	2.91
	Carvacrol	0.72	9.51	19.85	6.2

nd – not determined

Based on the SPME analysis, the emission of the compounds from the matrixes looked much less distinctive one from another if other absorbents (polar CW/DVB and polyacrylate) were used. In these tests, the amounts of analogous volatile compounds were of the same order for any oil above any matrix. On the other hand, the differences between the results with different absorbents was negligible if relative amount of the volatile compounds was analysed.

4 Discussions

4.1 Comparison of the matrixes
Compared as a whole, the Hi Cap100 was the least emitting matrix. The conclusion was supported by the results of the transient response and the SPME analysis. On the other hand, only part of the SPME results was coinciding with the conclusion. Depending on the absorbent, the SPME results supported the uniqueness of the Hi Cap100 (non-polar polydimethylsiloxane absorbent) or revealed a quantitative similarity between the compounds emitted from the matrixes. Though this difference between the absorbents was not clearly understood, the SPME based on non-polar polydimethylsiloxane was the method which results were compared with the characterisation of the matrixes by the 2D-τ-charts in our study. Consequently, some features defined in the 2D-τ-charts were related with certain compounds of the oils.

Based on the results in table 1, the cluster 3 in the chart Capsul E in figure 2 was supposed representing Limonene in the odour of oregano. In the charts Encapsul 855 and NLOK in figure 2, the clusters 1 and 3 could be related with the presence of Carvacrol above the matrix. It should be noted, that both more and less volatile compounds could generate short τ-components in the transient response which is depended mostly on the gas-surface interaction. The smallness of the weigh coefficients indicated small amount of the compound. Long τ-components marked by 1 in the charts Encapsul 855 and Capsul E in figure 3 seemed indicating the presence of Thymol in the air. The τ-component 2 in the chart Encapsul 855 in figure 3, could correspond with Thymol or p-Cymene. Unfortunately, the compounds not listed as the main in table 1 could induce the transient signals of the sensors. Therefore, the identity of specific clusters in the charts was not strict. On the other hand, a scattering of the points in the 2D-τ-charts could appear due to the limits of the approximation.

4.2 Limits of the transient analysis

It is clear that the main information in the transient is located in the initial part of it where the signal changes rapidly. In a long time domain the signal is mostly influenced by noise and its average value is nearly constant. In addition, some information could be lost because the transients had a fixed length in the experiments.

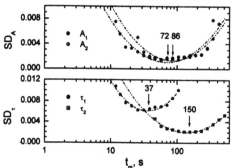

Figure 5. The relative standard deviation of the amplitudes and time constants as a function of the transient length: $A_1=A_2$ and $\tau_1=1$, $\tau_2=20$.

The optimal length is important for exponential approximation of this type of the transient signals. Strictly speaking even for a mono-exponential transients the single optimal length does not exist. It was found from the model calculations that in respect of the time constant the optimal length is 13τ, while in respect of amplitude roughly 20% shorter transients are optimal. This difference is even more noticeable when the transients composed of the components with large span in time constants are analysed. Typical calculations were illustrated for the two-exponent-transient in figure 5.

Since the experimental signals were usually interfered with a noise, the accuracy of the exponent approximation should evidently depend on the noise level. We performed statistical tests to determine the largest noise-to-signal ratio (NSR), at which the program reliably separates the components of the transient. Relative standard deviation of 10 % for time constant and 15% for amplitude calculation has been accepted as the criterion. Two and three-component transients were analysed. Having in mind that the typical NSR of the investigated transients was in the range 10^{-3}-10^{-2} we concluded from the tests that the integration method can be successfully applied for transient processing to distinguish two or even three components in experimentally measured transients.

References

[1]   Galdikas A, Mironas A, Senuliene D and Setkus A 2000 *Sensors and Actuators B* **68** (in press: SNB 3452).
[2]   Bakan J A 1973 *Food Technol.* **27** 11-44.
[3]   Tittelbach-Helmrich K 1993 *Meas. Sci. Technol.* **4** 1323-1329.
[4]   Press W H, Teukolsky S A, Vetterling W T and Flannery B P 1994 *Numerical Recipes in C* (Cambridge: University Press) p.120.

Section 2: Sensors/Instrumentation
Paper presented at the Seventh International Symposium on Olfaction and Electronic Noses, July 2000

49

Influence of Nitrogen Dioxide on the Response of Sputtered SnO$_2$ Thin Films to Carbon Monoxide

D G Rickerby

Institute for Health and Consumer Protection, European Commission Joint Research Centre, 21020 Ispra VA, Italy

M C Horrillo

Laboratorio de Sensores, Centro de Technologias Fisicas CSIC, Calle Serrano 144, 28006 Madrid, Spain

Abstract. The response of SnO$_2$ thin film gas sensors, fabricated by r.f. sputtering, to low concentrations of CO and CO-NO$_2$ mixtures in air is determined at temperatures in the range 150-350°C. The performance of these sensors is discussed in relation to the morphology and microstructure of the oxide thin film and the influence of a Pt layer on selectivity and sensitivity is investigated.

1. Introduction

Solid state chemical sensors, for measuring low concentrations of pollutant gases in air, are currently being developed because environmental protection legislation requires monitoring and control of potentially harmful gases in the atmosphere, generated by motor vehicle and industrial emissions.

Nitrogen dioxide and carbon monoxide represent a significant public health hazard. The first alarm threshold limits are set at concentrations of 15 mg/m^3 (13 ppm) for CO and 200 μg/m^3 (106 ppb) for NO$_2$ [1, 2].

Tin dioxide sensors are effective for detecting these gases because of their excellent response, operational simplicity, small size and low cost in comparison with other analysis techniques, such as gas chromatography, chemiluminescence and IR-absorption.

The operating principle of the sensors is based on the change in conductivity due to the chemisorption of gas molecules at the semiconductor surface. Depending on whether the reaction is oxidizing or reducing, acceptors or donors will be produced at the film surface, leading to the formation of a space-charge layer and modification of the free carrier density in the conduction band [3, 4].

2. Sensor fabrication and testing

The sensors were fabricated by depositing thin films of SnO_2 of 100-500 nm thickness with reactive r.f. magnetron sputtering from a 99.9% pure SnO_2 target on polycrystalline alumina substrates. Some of the sensors were in addition doped by sputtering a thin layer of Pt under an argon atmosphere on to the film. The presence of this catalyst improves the sensitivity and selectivity of the sensor response to CO and NO_2 [5-7]. Gold contacts were also deposited by sputtering.

The sensors were tested in a steel cell at a constant temperature and their resistance was determined at step concentrations of CO in air both with and without the presence of a small (0.1 ppm) interfering concentration of NO_2. Measurements were carried out over the temperature range 150-350°C, at CO concentrations from 5-100 ppm, in synthetic air. All sensors were subjected to a thermal annealing treatment at 375 °C for 12 hours, before the gas sensitivity measurements were performed.

3. Characterization of the films

The films were characterized by Glancing Angle X-ray Diffraction (GAXRD), Scanning Electron Microscopy (SEM) and Transmission Electron Microscopy (TEM) to determine their morphology, composition and microstructure.

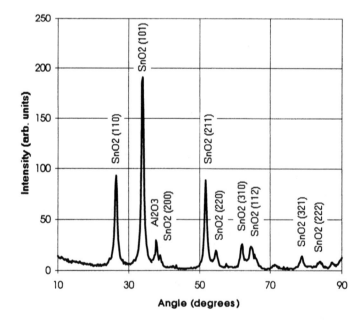

Fig. 1. GAXRD spectrum from a reactive r.f. sputtered SnO_2 film on an alumina substrate.

GAXRD indicated that the films had a polycrystalline structure without a preferential grain orientation (Fig.1). The angle of incidence for these measurements was 1°, leading to a calculated normal penetration depth of approximately 120 nm in the tin dioxide. The average diameter of the SnO_2 crystallites, estimated by applying Scherrer's equation to the diffraction peak broadening, was found to lie in the range 3 to 10 nm. Following the thermal treatment an increased number of SnO_2 peaks were apparent due to increased crystallinity of the film. The mean grain size did not however change significantly.

Figure 2 shows a high resolution scanning electron micrograph illustrating the surface morphology of the SnO_2 film. The nodular surface features are of dimensions comparable to the grain size of the polycrystalline alumina substrate. Granular features of ~10 nm diameter are also discernable on the surface of the film. The macroscopic roughness of the substrate, together with the nanocrystalline nature of the film, acts to increase the effective surface area of the oxide exposed to the gas [8].

Fig. 2. Field emission SEM image showing the surface structure of the SnO_2 film.

The nanocrystalline structure of the oxide film is also evidenced in the transmission electron micrograph shown in Fig. 3. The mean grain diameter is 12-15 nm [9], in reasonably close agreement with the X-ray measurements considering the approximations inherent in the latter. Sensitivity increases with decreasing grain size and is greatest when the grain diameter is approximately equal to or less than twice the Debye length, which determines the depth of the space charge layer [10, 11].

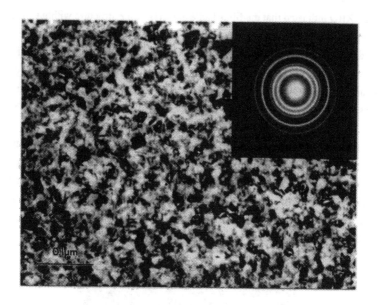

Fig. 3. Transmission electron micrograph of the film with corresponding electron diffraction pattern.

4. Sensor performance

The sensor response is defined as the ratio of the electrical resistance in NO_2-containing air to that in pure air, R_{NO2}/R_{air}, and viceversa for CO, R_{air}/R_{CO}, since the effect on resistance is opposite for oxidizing and reducing gases.

The optimum response for CO/air mixtures in the case of undoped SnO_2 was found to occur between 300-350°C, while below 250°C it was negligible [12]. The response was immediate and adsorption and desorption rates appeared similar for all CO concentrations measured.

Deposition of a 12 nm surface layer of Pt on the sensors increased their sensitivity and the optimum response was obtained at a lower operating temperature, 250°C, than for undoped SnO_2. For example, the response to 20 ppm of CO/air was 3.7 (Fig.4), compared to only 2.5 at 350°C for undoped films and high sensitivity was already attained at 150°C, as a result of the catalytic effect of Pt on the CO oxidation reaction.

For practical applications it is essential to understand the contribution of both gases to the overall sensor response. NO_2 is an electron acceptor, while CO is an electron donor. The effect of 0.1 ppm NO_2 on the response to CO of an undoped SnO_2 film is illustrated in Fig. 5. It is observed that at 350°C the interference effect is greatly reduced, while at 200°C the response to NO_2 dominates the behaviour. The sensor can therefore be made selective by varying its operating temperature.

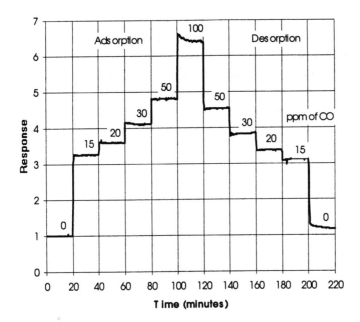

Fig. 4. Response to CO in air for a Pt doped SnO₂ sensor at 250°C.

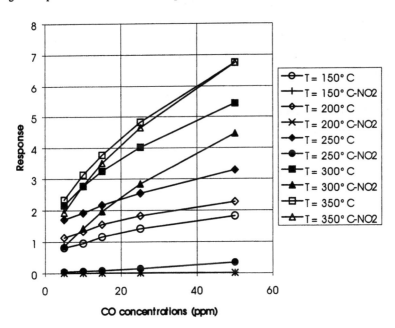

Fig. 5. Response to CO in air with and without 0.1 ppm of NO₂ of an undoped SnO₂ sensor.

5. Conclusions

Gas sensors of high sensitivity have been made by depositing ultra-fine grain size SnO_2 films on polycrystalline alumina substrates. The active surface area of the sensor is increased as a result of the substrate roughness and the nanocrystalline structure of the film. Deposition of a thin Pt layer increases the sensitivity and the capability to discriminate between NO_2 and CO. The selectivity can be further improved by varying the operating temperature. Sensors of this type are able to detect CO at the 1 ppm level and NO_2 at 10^{-1} ppm concentrations.

References

[1]   Gazzetta Ufficiale della Repubblica Italiana 1983 145 4209-4224

[2]   Official Journal of the European Communities 1985 28A L87/5

[3]   Chang S C 1974 IEEE Trans. Electron Devices 26 1875-1880

[4]   Göpel W and Schierbaum K D 1995 Sensors Actuators B26-27 1-12

[5]   Gutiérrez F J, Arés L, Robla J I, Horrillo M C, Sayago I, Getino J and Agapito J A 1993 Sensors Actuators B15-16 354-356

[6]   Sayago I, Gutiérrez J, Arés L, Robla J I, Horrillo M C, Getino J, Rino J. and Agapito J A 1995 Sensors Actuators B26-27 19-23

[7]   Schweizer-Berberich M, Zheng J G, Weimar U, Göpel W, Bârsan N, Pentia E and Tomescu A 1996 Sensors Actuators B31 71-75

[8]   Rickerby D G, Horrillo M C, Santos J P and Serrini P 1997 Nanostruct. Mater. 9 43-52

[9]   Xu C, Tamaki J, Miura N and Yamazoe N 1991 Sensors Actuators B3 147-155

[10]   Yamazoe N 1991 Sensors Actuators B5 7-19

[11]   Rickerby D G, Horrillo M C 1998 Nanostruct. Mater. 10 357-363

[12]   Barbi G B, Santos J, Serrini P, Gibson P N, Horrillo M C and Manes L 1995 Sensors Actuators B 24-25 559-563

Section 2: Sensors/Instrumentation
Paper presented at the Seventh International Symposium on Olfaction and Electronic Noses, July 2000

55

Development of Amperometric Gas Sensors for Detection of N₂O in Low Concentrations

Siswoyo[1], K C Persaud[1] and V R Phillips[2]

[1]*Department of Instrumentation and Analytical Science, UMIST, PO Box 88 Sackville Street, Manchester, M60 1QD, UK.*
[2]*Silsoe Research Institute, Wrest Park, Silsoe, Bedford, MK45 4HS, UK.*

Abstract. Two types of amperometric sensors for nitrous oxide gas have been investigated: unshielded electrodes and solid polymer bonded electrodes. Gold, silver and platinum were employed as working electrodes in the unshielded system. Both aqueous and non-aqueous solutions were used as electrolyte. The solid polymer bonded electrode has indicated to be a promising candidate for nitrous oxide gas sensors with increased sensitivity due to direct contact of the gas with the electrode.

1. Introduction

Nitrous oxide, N₂O, has been widely used as an anaesthetic and analgesic agent in clinical fields and as a propellant for pressurised containers in the food industry. It also has gained environmental importance as a greenhouse gas that may potentially destroy the ozone layer. It is very important, therefore, to monitor and control its concentration and emission. The concentration range of interest spans from part per billion to several percent.

The current methods of analysing N₂O include infrared spectrometry [1], gas chromatography [2] and refractometry [3]. Electrochemical methods have gained attention due to their simplicity, low cost and easy construction of devices compared with the other methods. Unfortunately, this kind of electrochemical apparatus is not available yet in the market. The existing electrochemical methods are currently based on reduction of nitrous oxide on silver, gold or platinum electrodes, employing both aqueous [4], and non-aqueous electrolytes [5]. Measurements are conducted with both unshielded and membrane covered electrodes. In the case of unshielded electrodes, a standard electrochemical cell is used without a shield between the electrolyte and the gas analyte. The range of the gas concentrations investigated are 0-100%. Limit of detection for N₂O has not been reported.

The main factor to be considered in the selection of N₂O sensors for measurements in ambient environments and agricultural fields is sensitivity. The existing electrochemical sensors use a gas permeable membrane to retain the electrolyte. As a result, the analyte gas has to pass the membrane first before reactions can take place at the electrode. Here, permeability of the membrane towards the analyte gas will affect the sensitivity, as well as the response time of the sensor.

The deposition of precious metals on solid polymer electrolytes (SPE) by chemical deposition allows the construction of efficient electrodes for the oxidation or reduction of gaseous analyte [6]. This approach led to the emergence of new types of amperometric gas sensors, where the working electrode is in direct contact with the gas phase. In this sensor design, the SPE separates the electrolyte compartment from the gas phase and simultaneously acts as an ion-conducting support for the working electrode facing the gas phase. The gaseous analyte is not required to diffuse through the membrane, as in the case of Clark electrodes, or to diffuse through a porous Teflon layer, as in the case of gas diffusion electrodes, prior to undergoing electron transfer [7,8].

This paper describes the development of N_2O sensors based on the SPE electrodes in comparison with the unshielded electrode configuration.

2. Experimental

2.1. *Unshielded electrode system*

Three metal electrodes were investigated for their responses to nitrous oxide: gold, platinum and silver. Gold and platinum wires (Aldrich) for working electrodes (WE) were used in a non-aqueous electrolyte, 0.2 M tetraethylammonium perchlorate (TEAP, Aldrich) in dimethylsulfoxide (DMSO, Fluka), in conjunction with a Ag/AgCl reference electrode (RE) and a platinum counter electrode (CE) in a glass cell. The sample nitrous oxide gas introduced into the cell was prepared using a simple technique. Pure N_2O gas (BOC Gas, 99.9%) and air (BOC Gas) were mixed in a glass container (100 ml). The flow rate of each gas was maintained at 20 ml/min. The concentration of N_2O was adjusted by varying its exposure time. Each measurement was carried out after stopping the gas flow into the cell. A Potentiostat/Galvanostat (EG&G Princeton Applied Research) based-computer was used for all measurements (Fig. 1). Cyclic voltammetric technique was applied between potential of 0–(–2.8) V vs. Ag/AgCl with a scan rate of 100 mV/s. A solution of 0.1M potassium chloride and 1M potassium hydroxide (KCl-KOH) was used as an electrolyte when silver wires were used as the working electrodes. The potential applied was 0 – (–1.6) V vs. Ag/AgCl with the same scan rate as before.

2.2. *Gold-Nafion electrode*

Deposition of gold onto Nafion®417 (Aldrich) surface was carried out using a chemical deposition technique [6,9]. Nafion®417 was initially boiled for 1 h in concentrated nitric acid (Fison Sci.) followed by similar treatment in distilled water for 1 h. A solution of 0.02M chloroauric acid (Fluka) and a solution of 0.05M sodium borohydride (Aldrich) in 1M sodium hydroxide (Fison Sci.) were used as the source of gold film and reductant, respectively. The deposition process was allowed to take place for about 24 hours at room temperature. The Au-Nafion membrane formed was then boiled in milliQ water for 1 h to release precipitates. This was followed by soaking the membrane in 0.5 sulfuric acid (Fison Sc.) for several hours, washing with water and drying in an oven. The resulting membrane was then used to construct the N_2O sensor (Fig.2). The electrolyte, counter and reference electrodes used were similar to the unshielded system. The electrodes were housed in a Teflon body (Omnifit). A piece of gold plate with a hole in the centre was used as a current collector and also to maintain a connection between the Au-Nafion working electrode and the potentiostat. Different concentrations of N_2O were then introduced into the sensor by varying the flow rate of N_2O (20, 40, 60, 80 and 100 ml/min) and holding the constant flow rate of pure air at 20 ml/min. Measurements were carried out under dynamic gas flow.

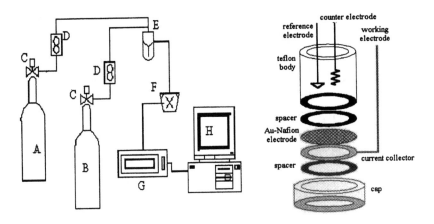

Figure 1. Experimental setup: A=air; B=N$_2$O; C=valve; D=flowmeter; E=mixer; F=cell electrode; G=potentiostat; H=computer.

Figure 2. Schematic of Au-Nafion electrode.

3. Results and discussion

3.1. *Unshielded Electrode*

From the results shown in Figs. 4 and 5, it is clear that the reduction of nitrous oxide on gold and platinum electrode using TEAP-DMSO as the electrolyte took place in the potential range -2.0 to –2.5 V vs. Ag/AgCl. In the case of silver electrode in KCl-KOH electrolyte, the reduction was observed between –1.1 and –1.5 V vs. Ag/AgCl (Fig. 3).

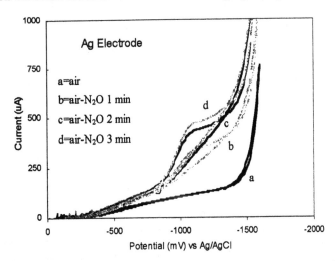

Figure 3. Cyclic voltammograms of Nitrous Oxide and Air at Ag wire electrode.

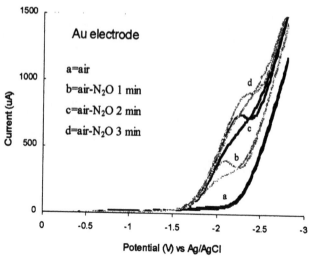

Figure 4. Cyclic voltammograms of Nitrous Oxide and Air at Au wire electrode.

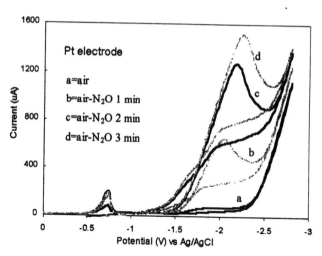

Figure 5. Cyclic voltammograms of Nitrous Oxide and Air at Pt wire electrode.

The current increased when the exposure time of nitrous oxide became longer. This is due to the fact that the longer the exposure time is the larger concentration of nitrous oxide in the solution, which in turn leads to a larger current produced due to a larger amount of nitrous oxide reduced.

When air alone was exposed to the electrode cell, the signal change was insignificant compared with that on exposure of the air and nitrous oxide mixture. Therefore, the interference from air was very small and this effect can be neglected compared to the signal obtained from N_2O gas. When the exposure time of gas was longer than 5 mins, there was no further current increase recorded, showing that the electrolyte has been saturated with nitrous oxide.

3.2. *Au-Nafion Electrode*

When the N_2O-air gas mixture was introduced onto the electrode, the response obtained was significantly different from that produced by air alone (Fig. 6). The increasing current was clearly recorded in the cyclic voltammogram. However, the response plateau as seen in the unshielded system (Fig. 4) was not clearly observed. This could be due to the quality of the Au-Nafion electrode itself or probably the difference in cell configurations employed in the two systems. A similar response has also been obtained when the electrode exposed to N_2 and N_2O-N_2 (Fig. 7). There was no current increase when the flow rate of N_2O was higher than 20 ml/min. This might be due to saturation of the sensor.

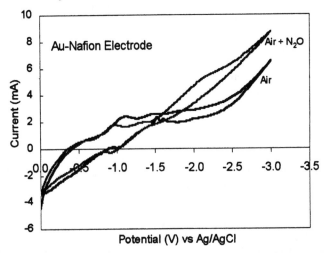

Figure 6. Cyclic voltammograms of Nitrous Oxide and Air at Au-Nafion electrode.

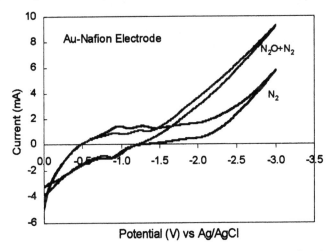

Figure 7. Cyclic voltammograms of Nitrous Oxide and Nitrogen at Au-Nafion electrode.

In order to know whether the Au-Nafion electrode is capable of detecting nitrous oxide gas in ppb range, more works should be carried out. Recent activities are mainly concerned with improvement of immobilisation process of gold onto the surface of Nafion, and investigation of sensor configuration. Investigation of platinum-Nafion electrode will be considered as an alternative because of the similarity of its electrochemical properties to gold. Although silver has proven its ability as a sensing material for nitrous oxide, it is less considered than gold and platinum because it is not inert enough when applied in the field.

4. Conclusion

Au-Nafion electrode can be a promising alternative for the nitrous oxide gas sensor. A quantitative characterisation is required in order to investigate the limit of detection and the working range concentration of the sensor.

Platinum and silver electrodes also give good response when applied in the unshielded cell configuration. It is possible to make Pt-Nafion and Ag-Nafion electrodes in the same procedure as alternatives to Au-Nafion electrode. Here, Pt-Nafion is more preferred because it is more inert than Ag, especially for application in outdoor environment.

References

[1] Schaffernicht H 1999 Drager Rev. 84 30-32.
[2] Kumagai S, and Koda S 1999 Am. Ind. Hyg. Ass. J. 60 458-462.
[3] Sugg B R, Palayiwa E, Davies W L, Jackson R, McGraghan T, Shadbolt P, Weller S J, and Hahn C E W 1988, Br.J.Anaesth. 61 484-491.
[4] Albery J W and Hahn C E W 1983 US Patent 4400242.
[5] McPeak H and Hahn C E W 1997 J.Electroanal. Chem. 427 179-188.
[6] Cook R L, MacDuff R C, and Sammells A F 1990 J. Electrochem. Soc. 137 187-189.
[7] Jordan L R, Hauser P C, and Dawson G A 1997 Anal. Chem. 69 558-562.
[8] Jacquinot P, Hodgson A W E, Müller B, Wehrli B, and Hauser P C 1999 Analyst 124, 871-876.
[9] Do J S, and Shieh R Y 1996 Sensors and Actuators B 37 19– 26.

Acknowledgement. The authors thank Silsoe Research Institute for the financial support. Siswoyo is supported by a scholarship from the DUE Project University of Jember Indonesia.

Section 2: Sensors/Instrumentation
Paper presented at the Seventh International Symposium on Olfaction and Electronic Noses, July 2000

61

Integrating Pervaporation with Electronic Nose for Monitoring the Muscatel Aroma Production

C. Pinheiro[1], T. Schäfer[1], C.M. Rodrigues[1], A. Barros[2], S. Rocha[2], I. Delgadillo[2], J.G. Crespo[1]

[1] Department of Chemistry, Faculdade de Ciências e Tecnologia, Universidade Nova de Lisboa, 2825-114 Caparica, Portugal
   Fax: + 351 21 2948385   email: carmen.pinheiro@dq.fct.unl.pt

[2] Department of Chemistry, Universidade de Aveiro, 3810-193 Aveiro, Portugal
   Fax: + 351 234 370084

Abstract. During the muscatel wine-must fermentation ethanol is produced in much higher quantities than aroma compounds. As a consequence, the electronic nose monitors mainly the evolution of the ethanol concentration, not perceiving the subtle variations of the aroma profile. This work introduces a membrane separation process - organophilic pervaporation - as a selective pre-enrichment step for the monitoring of the muscatel aroma production with an electronic nose.

1. Introduction

During the muscatel wine-must fermentation the aroma profile changes notably due to the yeast metabolism. The result is a very complex aroma composed of a multitude of compounds, many of which differing with regard to concentration, chemical and organoleptic properties, and contribution to the muscatel aroma. In the course of the fermentation these aroma compounds are produced in the ppm range, whilst one of the main metabolic products, ethanol, is produced in much higher quantities (up to about 10 wt%). For such a complex and valuable process it is important to have accurate and immediate information on the organoleptic development of the fermentation, such that it renders possible to detect and correct undesired deviations at an early stage. Traditionally the muscatel fermentation is monitored in the wine industry accompanying the evolution of the wine-must density along with human sensory analysis, and controlled solely by adjusting the temperature.

Monitoring the wine-must fermentation with an electronic nose can be of extreme value since the operating principle of this technique mimics the human olfactory system. The fast analysis time, simplicity of operation and low cost allows it to be implemented as an on-line, non-invasive, real-time and routine monitoring technique, thus enabling the automated control and optimisation of the fermentation. Due to the characteristics of the muscatel fermentation (production of ethanol in a high concentration along with aromas typically in the ppm range), however, an electronic nose might essentially monitor the increase of the ethanol concentration, not perceiving the subtle variations of the aroma profile. In this case, a selective pre-enrichment of the wine-must in aroma compounds relatively to ethanol may permit the electronic nose to subsequently discern the contribution of the aromas and be able to monitor the evolution of the aroma profile.

Organophilic pervaporation is a membrane separation process highly suitable for this selective pre-enrichment. It does not require any extraction aid, operates at ambient temperature, and can therefore be directly coupled to an active wine-must fermentation without deteriorating neither the fermentation nor the quality of the permeate obtained. In this process a hydrophobic dense membrane separates the upstream (feed) side containing the feed solution, from the downstream (permeate) side containing the compounds recovered. The transport of the solutes across the membrane is based on a gradient of the partial pressure of each compound between the liquid feed (at ambient pressure) and the vapour permeate (at a lower pressure). The hydrophobic membrane is not a mere barrier, but rather interacts with the solutes, promoting or rendering more difficult their sorption and subsequent diffusion according to their mutual chemical nature, i.e., it acts as a selective barrier. Therefore, due to the higher affinity of the membrane for the aroma compounds rather than ethanol, the permeate obtained is selectively enriched in aroma compounds and the original relation between ethanol and aromas in the feed solution (wine-must) shifted. The interaction between the membrane and the different wine-must aroma compounds permits to recover low and high volatile compounds equally well [1].

The work presented first investigates whether the electronic nose is able to monitor the aroma profile evolution along the fermentation or merely follows the production of ethanol. It then analyses if using organophilic pervaporation for selectively pre-enriching aroma compounds improves sample discrimination on the basis of the aroma profile.

2. Experimental

The muscatel wine-must was provided by a Portuguese wine-maker, José Maria da Fonseca and Sccrs. A 50L wine-must fermentor was inoculated with a commercially available strain of *Saccharomyces cerevisiae*. The fermentation process was followed for 20 days in terms of density of the wine-must, redox potential, number of viable and non-viable cells, glucose and fructose consumed, and aroma and ethanol content. The temperature was controlled at about 18 °C.

A commercial electronic nose (A32S, AromaScan, UK) with an array of 32 organic conducting polymers was applied. Samples were taken on-line directly from the fermentor headspace and also off-line from the fermentation broth. Off-line fermentation samples, as well as standards simulating the fermentation course, were equilibrated for 1 hour at 60 °C in a static-headspace multisampler system (A32S/50S, AromaScan, UK) prior to analysis. The on-line and off-line data were evaluated using a Principal Component Analysis software provided with the equipment (AromaScan, UK).

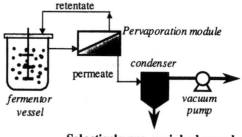

Figure 1. Laboratory-scale pervaporation unit

Selective sample pre-enrichment was carried out in a standard laboratory-scale pervaporation unit (Figure 1), using an organophilic POMS-PEI composite membrane (GKSS, Germany) [1].

3. Results and discussion

3.1. On-line and off-line monitoring of the muscatel wine-must fermentation

Figure 2 depicts results from the on-line monitoring of the fermentation with the electronic nose and Figure 3 from the off-line monitoring. The arrows in both figures indicate the evolution of the fermentation, with the wine-must density decreasing along the fermentation due to the production of ethanol. It can be observed that using the electronic nose and subsequent PCA analysis it was possible to visualise the fermentation evolution, with very similar trends during both the on-line and off-line measurements.

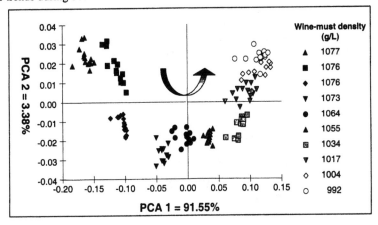

Figure 2. On-line monitoring of the muscatel wine-must fermentation.

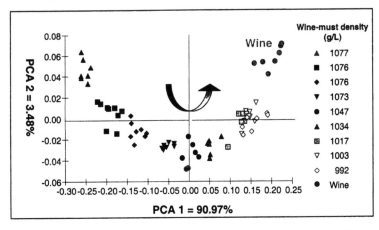

Figure 3.   Off-line monitoring of the muscatel wine-must fermentation. A muscatel wine from José Maria da Fonseca and Sccrs. (João Pires) was also analysed.

3.2. Monitoring the aroma profile or ethanol?

During the fermentation ethanol is produced in much higher quantities than aroma compounds. It was therefore investigated if the electronic nose was following the evolution of the aroma profile or simply the ethanol production.

Two sets of standards simulating the fermentation were analysed off-line so as to evaluate whether the electronic nose perceived the aromas produced. Eight aroma compounds were selected to mimic the muscatel wine-must, chosen either due to their relatively high concentration in the wine-must, or their organoleptic significance, or both [1]. The aroma compounds selected and their individual concentration range in the standards (similar to that observed during the muscatel fermentation) were: isoamyl alcohol (2 – 140 ppm); ethyl acetate (0.6 - 40 ppm); isobutyl alcohol (0.5 - 35 ppm); isoamyl acetate (0.06 – 4 ppm); ethyl hexanoate (0.05 – 3 ppm); hexyl acetate (0.05 – 3 ppm); 1-hexanol (0.02 – 1 ppm); and linalool (0.02 – 1 ppm). For simplicity, the sum of all aroma concentrations was considered in the following. The first set *A* consisted of seven standards with simultaneously increasing aroma and ethanol concentrations; the second set *E* did not contain aromas and consisted of seven standards with increasing concentrations of ethanol only. Hence, the samples in the two sets *A* and *E* could be grouped two by two according to the identical ethanol content, with the difference that *A* additionally included aroma compounds (table I).

Table 1: Aroma and ethanol concentrations in the standards of sets A and E, as well as the selectively pre-enriched sample.

	Ethanol (g/L)		Ethanol (g/L)	Aromas (mg/L)
E1	1.56	**A1**	1.56	4.60
E2	3.13	**A2**	3.12	9.20
E3	6.25	**A3**	6.23	18.39
E4	12.50	**A4**	12.46	36.79
E5	25.00	**A5**	24.92	55.94
	Selectively Pre-enriched		24.92	809.00
E6	50.00	**A6**	49.85	147.16
E7	100.00	**A7**	99.71	294.31

These standards where analysed off-line with the electronic nose. Figure 4 presents the PCA plot of the two sets (open symbols refer to samples *E* and filled symbols to samples *A*).

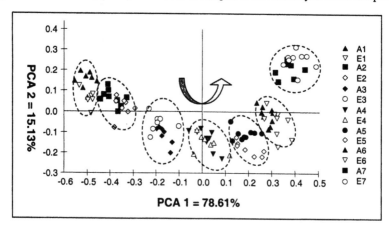

Figure 4. Standards simulating the fermentation evolution in terms of aroma compounds and ethanol (set *A*), and ethanol only (set *E*).

It can be observed that using all the 32 sensors in the array the electronic nose could not discriminate between *A* and *E* clusters of the same ethanol content, with and without aromas. Selection of individual sensors did not improve this discrimination because not any of the 32 sensors yielded a different response to standards with and without aromas. Hence, under the experimental conditions the presence of aroma compounds in the standards did not cause any significant response in the sensors, which were consequently responding to ethanol only.

3.3. Selective pre-enrichment using pervaporation

It was then analysed if by using pervaporation for increasing the aroma concentration relatively to ethanol the electronic nose could detect the contribution of the aromas.

One of the A samples (*A5*), simulating a highly aroma productive period of the fermentation with about 25 g/L ethanol and 56 mg/L total aroma concentration, was pervaporated. The permeate was quantitatively recovered and then diluted to the initial ethanol content (25 g/L), nevertheless yielding a total aroma concentration 14 times that of the original (809 mg/L, Table I).

Figure 5 depicts the PCA plot of the enriched sample together with the two sets of samples *A* and *E* from Figure 4. It can be seen that the selectively pre-enriched sample deviates strongly from the tendency of the sets *A* and *E* predominantly ruled by the increase of the ethanol concentration. It is very well discriminated from the original *A5* and corresponding *E5* sample, which can only be attributed to the aroma compounds because the ethanol content is very similar in all three samples.

Consequently, despite of high ethanol concentrations, integrating pervaporation as a selective pre-enrichment step the electronic nose can distinguish samples on the basis of the aromas, and it may permit to monitor the variations of the aroma profile along the muscatel wine-must fermentation.

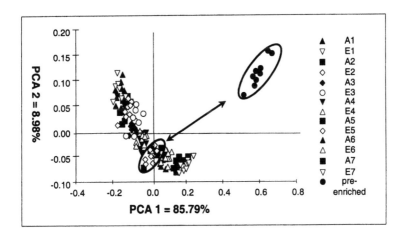

Figure 5. Comparison of a selectively pre-enriched sample using pervaporation and standards simulating the fermentation evolution in terms of aromas plus ethanol (set *A*) and ethanol only (set *E*).

4. Conclusions and Future Work

The results obtained show that without sample pre-treatment the electronic nose could only perceive the evolution of ethanol during the muscatel fermentation, but not the subtle variations of the aroma profile.

However, we have demonstrated the potential of integrating pervaporation as a selective pre-enrichment step with the electronic nose in order to monitor the aroma contribution in samples where ethanol is in much higher concentrations. Future work will evaluate the feasibility of using this integrated system to monitor the aroma profile along the muscatel wine-must fermentation. This selective pre-enrichment should, however, not impair the main advantages of the electronic nose, namely the simplicity of operation and the short analysis time. Therefore, a fast, simple and automated unit of integrated Pervaporation-Electronic Nose should be established.

References

[1] Schäfer, T., Bengtson, G., Pingel, H., Böddeker, K.W., Crespo, J.P.S.G. 1999. "Recovery of Aroma Compounds from Wine-Must Fermentation by Organophilic Pervaporation". Biotechnology and Bioengineering, Vol.62, No.4, February 20: 412-421

Section 2: Sensors/Instrumentation
Paper presented at the Seventh International Symposium on Olfaction and Electronic Noses, July 2000

67

Detection of TNT vapours with the Pico-1 Nose

M. Pardo[1], G.P. Benussi, G. Niederjaufner, G. Faglia and G. Sberveglieri

INFM and Gas Sensor Lab, Dept. of Chemistry and Physics for Materials
University of Brescia, Via Valotti 9 - 25133 Brescia -Italy

Abstract. The problem of sensing military grade TNT is addressed with the Pico-1 Electronic Nose developed at the Brescia University. Pico-1 is based on semiconductor thin films and uses static headspace sampling. TNT in air, sand and soil substrates is measured. All measurements were carried out against their respective substrates. It is found out that the TNT detection can be easily performed in two steps. First a Multilayer Perceptron (MLP), using as inputs the features extracted from the responses of three SnO2 catalyzed sensors, distinguishes between the different substrate types, irrespectively of the presence of TNT. Having determined the substrate, two bare sensors suffice to check if TNT is present: PCA plots show that the measurements quite clearly form two clusters (TNT absent/present). Further, regarding the detection limit of single sensors for TNT in air, measurements performed at room temperature (4 ppb of saturated TNT concentration) showed that there is a 10% variation of the sensor response (with respect to air) when TNT is present.

1. Introduction

The detection of unexploded ordnance such as antipersonnel mines is presently a lengthy and dangerous procedure. This task is carried out with the help of metal detectors or trained dogs. To render the detection more efficient various techniques are being proposed such as infrared spectrometry, radar and multispectral imagery analysis [1].

Recently, the proposal of using chemical sensors to detect vapors emitted from explosives has been put forward. DNT (2,4 dinitrotoluene) in the ppb range was sensed by SAW devices coated with polymers [2] and both DNT and TNT were detected measuring the photoluminescence spectrum of Porous Silicon [3]. At the present level of technology no single detection method can alone serve the purpose; data fusion seems to be mandatory for real applications.

[1] E-mail : pardo@tflab.ing.unibs.it.

In the last decade, the use of arrays of sensors in stand-alone devices for the detection and classification of odors has given rise to the field of Electronic Noses (EN). ENs recognize a global information (fingerprint) of the volatile compounds in contrast to classical techniques such as gas-chromatography in which the single components are identified. They consist of a sampling system, electronic circuitry, an array of sensors and data analysis software. ENs have been mainly applied to quality monitoring (foods, packages) (see e.g. [4]).

Given the small TNT vapor pressure, the detection mechanism is probably indirect: DNT which is present in low concentrations in the TNT has a much higher vapor pressure and is therefore present in the gas phase in greater concentration.

Table 1. The sample sequence

Vial number	Vial content
1	0.3g TNT
2	Air
3	0.5g TNT + 2.5g soil
4	2.5g soil
5	0.6g TNT + 6g soil
6	6g soil
7	0.5g TNT + 2.5g sand
8	2.5g sand
9	0.45g TNT
10	Air

2. Experimental

2.1. Sensors

Five SnO_2-based sensors (two pure SnO_2 and three SnO_2 catalysed with Pd, Au and Pt) were deposited in a DC magnetron sputtering system from a pure Sn target according to the RGTO technique [5]. The RGTO (Rheotaxial Growth and Thermal Oxidation) technique consists of two steps: first the deposition a metallic film with DC magnetron sputtering from a tin target on a substrate, kept at a temperature higher than the melting point of the material, then the thermal oxidation in humid air in order to get a stoichiometric composition for the metal. The deposition of the metal on a substrate kept at temperature higher than the melting point of the metal (350C) makes the film to grow with a high surface area. The thickness of the metallic film deposited is circa 300 nm for our samples. SEM micrographs of the surface after the deposition show that there are spheres with dimensions ranging from 1 mm to 3 mm that aren't interconnected, hence there is no current flow between the contacts. The metal-semiconductor phase transformation (Thermal Oxidation) is achieved by keeping the thin film in a furnace with a humid synthetic air flow (200 l/min) and making a thermal cycle of four hours at 250C and thirty hours at 600C. During these hours a little layer of oxide grow on the

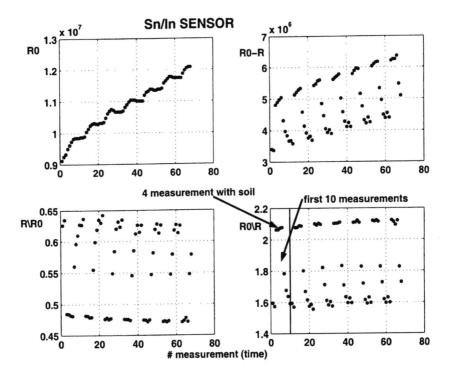

Figure 1. Indium catalyzed sensor, one point per response curve. From top left clockwise: R_0, $R - R_0$, $\frac{R_0}{R}$, $\frac{R}{R_0}$. The 4 measurements with soil in the $\frac{R_0}{R}$ plot are not distinguishable, while the various substrates are distinguishable.

surfaces of the spheres and the second step at 600C complete the oxidation. The oxygen incorporation in the lattice increases the volume of about the thirty percent, causing interconnection of the agglomerates and creating percolating paths for the current flow. SEM images of the surface of the semiconductor film after thermal oxidation show that the agglomerates are porous. This surface configuration has been proven to be well suited for gas absorption. To improve the selectivity of the sensors, we deposited a thin layer of noble metals (Palladium, Platinum and Gold) as catalyst on three sensors. The thickness of these layers, deposited by sputtering onto the tin oxide surface at 300C for Platinum and 200C for the others, was 4 and 5 nm respectively. TEM analysis shows that catalyzers form nanometric clusters.

2.2. Sampling

Each sample shown in table 1 is introduced into a vial with a volume of $20cm^3$ which is crimped with seal and septa. The vial is left in an incubation oven at 40C for 30 minutes

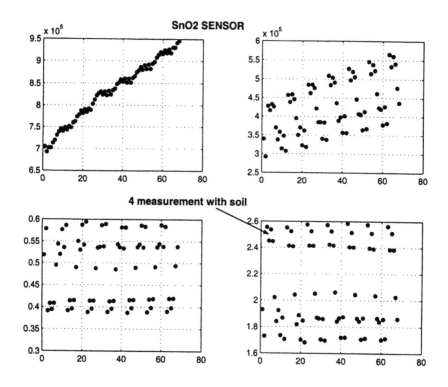

Figure 2. Pure sensor, one point per response curve. From top left clockwise: R_0, $R-R_0$, $\frac{R_0}{R}$, $\frac{R}{R_0}$. The measurements with soil in the $\frac{R_0}{R}$ plot form 2 clusters corresponding to TNT present/absent.

in order to reach gas-solid equilibrium. Part of the headspace is then extracted with a syringe and injected, at strictly constant velocity, in the air flow which is used as carrier (static headspace scheme).

Six successive extractions of the sequence shown in table 1 were performed. The extractions were separated by a time interval sufficient to reach equilibrium. As shown in table 1 all measurements were carried out against their respective substrates. In this way we check that the different responses are not due entirely to the different substrate (though this is the main cause). The sand samples are obtained by grinding pebbles for the manufacturing of concrete. Since solid TNT in equilibrium is always present, all measurements refer to a saturation condition of the gas phase. This equilibrium is influenced by the substrate. In particular we expected similar results for the measurements with sand and with air, since sand was thought to be inert, and a smaller response for measurements with soil because of the interaction of TNT with the soil.

Furthermore the response to TNT in air held at different temperatures (23, 40 and

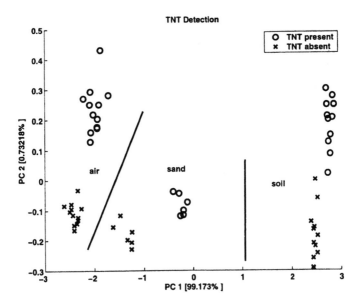

Figure 3. PCA plot with all sensors.

50C), and therefore at different saturation concentrations, was analyzed. 23C correspond to circa 4ppb of TNT in the headspace.

2.3. Data Analysis

For each response curve we preliminarily extracted one point of the baseline and three tentative features, $R-R0$, $R/R0$ and $R0/R$ (R steady state resistance, $R0$ baseline resistance). The time evolution of these features were then plotted to see the drift and to have a first impression of the classification ability of each sensor. These four plots are shown in figure 1 for the Indium-catalysed sensor and in figure 2 for a pure sensor. This exploratory analysis permitted to draw the following considerations:

1. Although a drift of the baseline is present the feature $R/R0$ doesn't drift. Therefore $R/R0$ is used as feature for every response curve.

2. The catalyzed sensors do not distinguish between TNT + substrate and substrate alone (with the exception of the sand samples), see the first ten points in the fourth plot in fig. 1. These points represent the first measurements sequence as described in table 1. As the sequence is repeated the measurements referring to a particular substrate remain on the same horizontal line, that is the have the same value.

3. The bare sensors do distinguish between TNT + substrate and substrate alone but this difference is less than the difference between the various substrates, see the

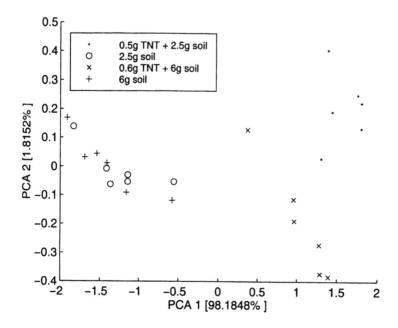

Figure 4. Separation between soil + TNT and soil.

difference in the soil samples as shown by the arrow in fig. 2.

The five sensors together give the PCA plot shown in figure 2. It is possible to detect the presence of TNT in two ways: either with a direct classification of the data in two classes (TNT present/absent) or by dividing this task in two steps. In the first case, a nonlinear separation technique has to be adopted. We choose multilayer perceptrons. PCA is used to reduce the input features removing the noise [6]. The inputs are then the projection of the data on the first principal components. The net was optimised with respect to the number of input neurons. To avoid overfitting we trained the nets both with early stopping and Bayesian regularisation (penalisation) [7]. A 2-5-2 gave 100% recognition on the training (48 points) and test data (12 points).

The second approach, which permits a visual classification and is more robust, is structured as follows:

1. Use the three catalyzed sensors to distinguish between the different substrate types, independently of the presence of TNT. Three distinct clusters can be recognized in fig. 3 even if, as expected, there isn't much distance between sand and air. The clusters are linearly separable. A 2-3-3 feedforward neural net was used to classify the data considering the scores of the two first PCs as inputs. 16 test patterns were correctly classified.

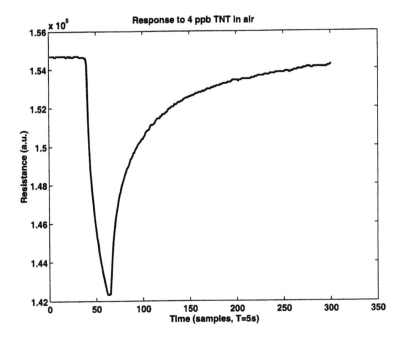

Figure 5. Response to 4ppb TNT in air.

2. Check with the two bare sensors if TNT is present. A distinction in two clusters, i.e. TNT + substrate and substrate alone, for every fixed substrate type is evident from the PCA plots. See fig. 4 for the case of soil, where a further distinction between the two different quantities of TNT is possible (samples 3 and 5).

Regarding the detection limit, we found out that even at room temperature (corresponding to a concentration of 4 ppb) the difference between the sensor responses with and without TNT is detectable (10% variation in the response). In fig. 5 we can also see that the response time is of the order of 100 s and that the sensor returns to the baseline value.

3. Conclusions

We showed that the problem of sensing military grade TNT in air, sand and soil substrates can be addressed by using tin dioxide thin film sensors and pattern recognition techniques. The sensors are capable of detecting TNT concentrations in the ppb region. Further work remains to be done regarding e.g. the reproducibility of the results between nominally equally prepared samples. Pico-1, given its high response time, is promising for on board use on a robot dedicated to demining.

References

[1] Miao X, Azimi-Sadjadi M, Tian B, Dubey A and Witherspoon N 1998 *IEEE Trans. Neur. Netw.* **9** 5

[2] Mslna T, Mowery R, and McGill R 1998 *Proc. Int. Meeting on Chemical Sensors, Bejing*

[3] Letant S, Content S, Zenhausern F, Tan T and Sailor M 2000 *Proc. Int. Conf. on Porous Semiconductors, Madrid* (94)

[4] Pardo M, Niederjaufner G, Benussi G, Comini E, Faglia G, Sberveglieri G, Holmberg M and Lundstrom I *Sensors and Actuators B* (to be published)

[5] Sberveglieri G, Faglia G, Groppelli S, Nelli P and Camanzi A 1990 *Semicond. Sci. Technol.* **5-41** 1231

[6] Bishop C M 1995 *Neural networks for pattern recognition* (Oxford University Press)

[7] Cherkassky V, and Mulier F 1998 *Learning from data* (Wiley)

Section 2: Sensors/Instrumentation
Paper presented at the Seventh International Symposium on Olfaction and Electronic Noses, July 2000

75

Design and Development of an Electronic Nose System to Control the Processing of Dry-Cured Iberian Hams Monitored via Internet

M. C. Horrillo, I. Sayago, M.J. Fernández, R. Gómez-Espinosa, A. Blanco, L. Otero, M. García, L. Arés and J. Gutiérrez.

Instituto de Física Aplicada (IFA), Consejo Superior de Investigaciones Científicas (CSIC), Serrano, 144, E-28006, Madrid, Spain.

Abstract. We present the design and development of an electronic nose system to control the processing and final product of dry-cured Iberian ham. Such novel system is able to control, acquire and transmit the data via Internet. Good results of response from a multisensor with eight semiconductor sensors for dry-cured Iberian hams that defer in their drying process, are also presented for different operating temperatures, observing differences ones.

1. Introduction

Due to recent advances in different interdisciplinary matters, such as organic and inorganic chemistry, sensor technology, electronics and artificial intelligence, the fabrication of electronic noses for the characterisation of aromas has been possible and it has become a commercial reality [1,2]. Along this decade this technology will find a wide range of commercial applications, such as medical diagnostics, environmental monitoring, and processing and quality control of foods. Within the former application it is the aroma of Iberian dry-cured ham, which is one of the most important characteristics that influences the acceptance of this product by consumers. The study of dry-cured ham flavour is very interesting to understand the formation of odorous compounds during the curing process. The knowledge of these compounds in each stage is very important for the food industry so that the process to obtain the final product can be controlled.

Until now, there have been few studies on the aroma components of hams and most studies have been carried out with sophisticated and expensive instrumentation such as gas-phase chromatography-mass spectrometry and extraction techniques such as high vacuum distillation and dynamic headspace. Such procedures have allowed to identify seventy-seven components in the ham flavour, most of which were alkanes, aldehydes, aliphatic alcohols, ketones and esters [3-5]. The great shortcoming of these techniques is that they are not able to measure the compounds in real time and in continuous, mainly during the stages of elaboration of the product. Furthermore, analysis time is very long, so that they can not be used routinely in the food industry.

In this paper we present an electronic nose (chemical sensor array) which is able to perform measurements in real time, in continuous, and in a fast way, so that direct analysis of complex gaseous mixtures (aroma) can be made [6,7]. The response of each array sensor is related to the whole of the species present in the environment of measurement due to its low selectivity. Therefore the sensor array system uses the global information formed for all the responses to classify the environment, and it extracts the information based on the requirement that each sensor must have different response values. In our case the obtained signal from each sensor is transmitted directly via Internet to anywhere to be evaluated later by pattern recognition methods, so that selectivity and discrimination capability can be improved. Thus, we can obtain very rapid diagnoses for the quality control of the product.

2. Experimental

Each sensor is distributed in radial way on an alumina substrate of 1 inch of diameter. Two semiconductor oxides, tin oxide and titanium oxide, were grown by sputtering (radio frequency type and magnetron mode) at 250°C using SnO_2 and TiO_2 targets under a 10% oxygen-90 % argon mixture at a total pressure of 0.5 Pa. Some SnO_2 sensors were doped with two different amounts of Pt and Pd by changing the deposition time during the process of sputtering. These dopants were introduced from metallic targets as an intermediate discontinuous layer between two layers with the same thickness. Electrical contacts were also deposited by sputtering. Table 1 shows the multisensor distribution used.

Table 1. Sensor array

Sensor	Material
1	SnO_2 (200 nm)
2	SnO_2 (400 nm)
3	SnO_2 (150 nm) + Pt (4 s) + SnO_2 (150 nm)
4	SnO_2 (150 nm) + Pt (8 s) + SnO_2 (150 nm)
5	SnO_2 (150 nm) + Pd (4 s) + SnO_2 (150 nm)
6	SnO_2 (150 nm) + Pd (8 s) + SnO_2 (150 nm)
7	SnO_2 (150 nm) + TiO_2 (100 nm)
8	SnO_2 (150 nm) + TiO_2 (150 nm)

The aroma is taken in two different ways:

1- Extraction from chamber with Iberian hams:

The extraction and measurement system of aroma from drying-chambers of Iberian ham is shown in Fig. 1. This electronic system carries out the control, measurement, and data acquisition and transmission via Internet. The system consists of: a pump which extracts the aroma coming from the drying-chambers to introduce it in the measurement cell where the multisensor (8-16 semiconductor-sensors) is installed. Since the aroma of the chambers is controlled by defined characteristics of humidity and temperature (temperature range: 10-30°C and relative humidity: 70-80% for the 14 months of ripening), we also control such parameters, so that the aroma (sample) is in the same conditions when it is introduced in the measurement cell. In this way the volatile organic compounds do not change with respect to those of the drying- chambers. The relative humidity is measured before the aroma enters in the measurement cell, so that the same conditions can be produced with our humidity system by mixing dry air with bubbling wet air. (Fig. 1). Then it will be introduced in the measurement cell to obtain the reference base line before introducing the aroma with that same humidity. Mass flow controllers (Fig. 1) monitor aroma flow as well as humid flow with a flow rate of 200 ml/min for both cases.

2- Extraction from final product in laboratory

5 g of crushed ham were placed in a Drechsel bottle and it was kept for 15 minutes to concentrate the aroma. Then the aroma was forced by a synthetic air flow of 200 ml/min into the measurement cell (Fig.2).

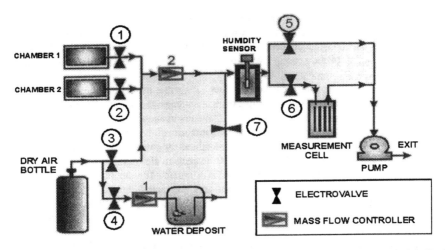

Fig. 1: Electronic system for measurement of aroma coming from chambers with Iberian hams.

Fig. 2: Measurement system of aroma from final product.

3. Software of the experimental system.

Hardware to process aromas must be compensated with an appropriate software which controls and supervises experiments, and collects, pre-processes and exports results, namely it allows a correct interaction between sensors and data supply. Moreover, data must be compatible with any standard application for its storage, visualisation, and processing. Two applications must be developed to carry out all data handling.

1. Control software

The used control software is LabView 5.1.1 running on an external PC, which is equipped with a GPIB Interface Card and a Multifunction I/O Board for PCI.

The extraction of aroma from the chambers is performed by a network that consists of pump, electrovalves and mass flow controllers, which are managed by the software. This also controls the function of each element at each step of the measurement cycles. LabView transmits requirements of the process cycle to the I/O board and then sends the appropriate electrical signals to control the power boards.

The main task is to collect data (resistance values) about the aroma that the multisensor is detecting, so that a multimeter and a multiplexer that switches the output of the measurement cell to the sensor resistances (8-16), are needed. LabView manages both instruments through the GPIB. On the other hand, the operating temperature of the multisensor is monitored through a PID temperature controller, which is regulated by internal programmation and its on/off status is also managed by LabView. With respect to the humidity measurement, an electric value is given by the humidity sensor and it is collected by the I/O board, which then compares it to the introduced reference values and the actual humidity is obtained by regulating air flows. The flow of the chambers-aroma, the flow of dry air to be mixed with wet air and the cleaning of the system are regulated by mass flow controllers connected to the I/O board, which is controlled by LabView, so that the required flow can be obtained at each moment. Figure 3 shows the general overview of control, measurement and data acquisition and transmission from chamber with hams.

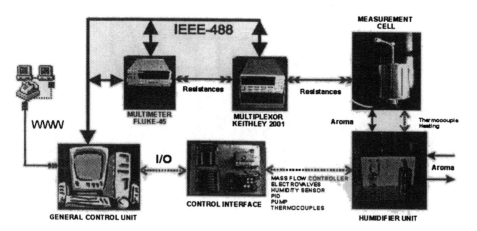

Fig. 3: General diagram of control, measurement, data acquisition and transmission from chamber with hams.

A data base is generated with all parameters of the experiment by means of a correct sequence of commands related to gas handling, flow control, cleaning cycles and reading and processing of results. A series of functions allows data filtering and handling as well as its three-dimensional presentation. Finally, an interface with other applications (Matlab, Excel, Access, etc.) is created for subsequent data processing.

2. Communication software
Its function is to establish an informative connection between the external environment and the experimental system. To accomplish this the use of protocols and applications universally established has been intended and Internet has been found as the most appropriate with precisely the World Wide Web, as one of its most spread protocols. Using web navigators as Netscape or Explorer yields advantages, such as that the experiment can be observed and/or

managed from any geographical location by any authorised person. Therefore it presents the advance of a telecontrol what can be very helpful in numerous experiences in laboratory, as well as in industry. It is possible to see in Fig. 4 how an experiment control is performed.

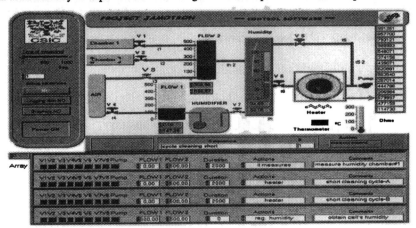

Fig.4: View of control panel in the Web page.

A"client-server" architecture has been used to implement the control, utilising mixed programmation: HTML-dynamic, Java Scripts and CGIs with programmation in G. In this way an optimigsed Web page type has been created with a changing content in manner selective. The result is that it is possible to control the experiment in real time, to access data and to transmit results to anywhere.

4. Preliminary results

Until now there are only results with the ready-to-eat ham. Two types of dry-cured ham have been tested, according to the highest temperature (20°C or 30°C) during the period in dryer (3-4 months), and different responses have been obtained, so that it has been possible to elucidate them. In the case of the dry-cured ham at lowest temperature it is observed that the equilibrium is worse (Fig. 5). This can be so because the increase of the operating temperature yields more oxidant compounds than in the case of the hams dried at the highest temperature.

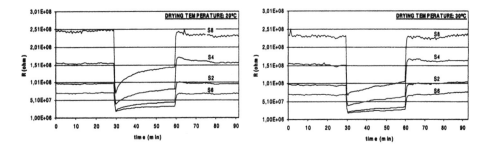

Fig.5: Different responses at 250°C for Iberian ham dried at two different temperatures.

Moreover it has been seen that, for the tested operating temperatures: 150-200-250°C, the best response is obtained at 250°C although at 150°C it is also possible a good detection of such compounds (Fig. 6). The performance is worse at 200°C because the semiconductor oxide resistance has a minimum at that temperature in air. Each sensor shows a different response, which will help to obtain a better discrimination of compounds on the site where the transmitted data will be handled afterwards.

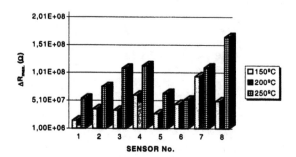

Fig.6: Responses to Iberian ham at different operating temperatures.

5. Conclusion

An electronic nose monitored via Internet has been developed to control the drying process of dry-cured Iberian hams in conditioned chambers. The system is versatile enough as to be used in any application for foods and also to combine different operating principles of sensors in order to improve the discrimination capability of a sensor array.

Furthermore, it has been possible to obtain a good response with a semiconductor multisensor of eight sensor elements to aromas coming from dry-cured Iberian ham and to elucidate two types of Iberian ham dried at different temperatures.

References

[1] J. Ouellette, "Electronic noses sniff out new markets", *The industrial Physicist*, No. Feb., pp. 26-29, 1999.

[2] T. T. Tan, S. Labréche and H. Amine, "Quality control of coffee using the FOX4000 electronic nose", ISIE' 97, Guimaraes, Portugal, 7-11 July 1997.

[3] C. García, J. J. Berdagué, T. Antequera, C. López-Bote, J. J. Córdoba, J. Ventanas, "Volatile components of dry cured Iberian ham", *Food Chemistry*, vol. 41, pp. 23-32, 1991.

[4] E. Sabio, M. C. Vidal-Aragón, M. J. Bernalte, J. L. Gata, "Volatile compounds present in six types of dry-cured ham from south European countries" *Food Chemistry*, vol.61, No 4, pp. 493-503, 1998.

[5] G. Barbieri, L. Bolzoni, G. Parolari, R. Virgili, R. Buttini, M. Careri, A. Mangia, "Flavor compounds of dry-cured ham", *J. Agri. Food Chem*, vol. 40, pp. 2389-2394, 1992.

[6] M. C. Horrillo, J. Getino, J. Gutiérrez, L. Arés, J. Robla, I. Sayago, J. Gutiérrez, "Measurements of VOCs with a semiconductor electronic nose", *J. Electrochem. Soc.*, vol.145, No 7, pp.2486-2489,1998.

[7] I. Sayago, M. C. Horrillo, J. Getino, J. Gutiérrez, L. Arés, J. I. Robla, M. J. Fernández, J. Rodrigo, "Discrimination of grape juice and fermented wine using a tin oxide multisensor", *Sensors and Actuators B*, vol. 57, pp. 249-254, 1999.

Section 3

Data Processing

Section 3: Data Processing
Paper presented at the Seventh International Symposium on Olfaction and Electronic Noses, July 2000

83

Methods for Sensors Selection in Pattern Recognition

A. Pardo, S. Marco, C. Calaza, A. Ortega, A. Perera, T. Šundic, J. Samitier

Instrumentation and Communication Lab. Departament d'Electrònica
University of Barcelona,
Martí Franques 1, 08028 Barcelona (Spain)
e-mail: toni@el.ub.es   http://nun97.el.ub.es

Abstract: Fast development in sensor technologies provides users a wide sensor spectrum for their applications. On the other hand, the selection of the optimum sensors among this wide spectrum can be a problem and a systematic method is needed. In this paper we present different methods to select sensors. A method based on Principal Component Analysis and Linear Discriminant Analysis are evaluated and they are compared with search methods as the classical Sequential Forward Selection (SFS) and Sequential Backguard Selection (SBS). Finally we have also studied Genetic Algorithms optimisers (GA).

1. Introduction.

Nowadays, advances of technology generate new sensors tuned to different gases of interest, although they are still non-selective. Many sensors from the same family can be found although their responses are only slightly different. This wide spectrum of sensor types is needed, but it generates a selection problem when a specific application has to be developed. Economic and/or technological aspects are also points to consider. Many papers are focused on sensor technology development or on powerful classification algorithms but often, the sensor selection intermediate step is forgotten. However it is well known that in most cases, a subset of sensors provides better recognition rate that the whole set of sensors and therefore the application can be simplified: using simpler classifiers and/or using less sensors. In addition, every extra sensor generates an additional cost and, moreover, the less sensors are used, the less training points are needed for the classifier. For these reasons, the number of sensors used for a certain application should be minimised.

Hence, a systematic method for selecting sensors in order to optimise a sensor array would be desired.

This work presents different methods for selecting sensors, that will be applied to problem of discerning gases with a group of twelve tin oxide sensors (from FIS, Osaka). The manufacturer has kindly provided the data.

2. Feature selection theory

The problem of the selection of the best sensors arises from the "curse of dimensionality": most of the classifiers do not provide the best result if irrelevant or redundant sensors are used [1].

This problem is usually prevented using methods of dimensionality reduction: feature extraction or feature selection.

With the feature extraction method, a mathematical transformation is implemented. This transformation maps the original sensor space to a new space, usually of lower dimensionality where the performance of the system improves. Among these transformations, linear methods as Principal Component Analysis (PCA) and Lineal Discriminant Analysis (LDA) are the usual choice.

In the case of feature selection, just a subset of features (sensors) of the original space is selected. This selection is guided by the improvement of a certain function, usually the recognition rate.

Among the different algorithms for feature selection, in this paper we have considered two main groups: methods based on lineal projection and search methods.

3. The problem

In this work we have used seven gases (Hydrogen (1-30ppm), carbon monoxide (1-30ppm), methane (10-300ppm), ethanol (1-30 ppm), Acetone (1-30 ppm), Acetic acid (1-30 ppm), and Toluene (1-30 ppm)) and the final aim has been to choose the best sensor subset for the classification problem. The entire sensor set contains twelve sensors from FIS – Osaka. Eleven of these sensors work in isothermal mode and one (sb-95) works in pulsed mode. The sensor list appears in figure 1.

From the complete sensor response, the steady state for 4 concentrations into the range described, has been selected. The feature vector has been built from the steady state sensor signals as given by:

$$S_i = \frac{1 - R_i / R_{o,i}}{\sum_i \left(1 - R_i / R_{o,i}\right)} \qquad (1)$$

where $R_{o,i}$ is the i sensor resistance in air.

For the classifier the K-nearest neighbours (K-NN) algorithm with K=1 has been used. Due to the scarcity of data points, validation is based on leave one out.

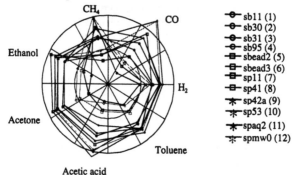

Figure 1: sensor response radar plot to maximum concentration range

With the complete set of twelve sensors, the number of possible subsets is 4095. The baseline result is the recognition rate achieved with the complete set of sensors: a 57%. Figure 1 shows the response of every sensor to the seven gases for the maximum concentration. From this figure is impossible to extract any rule for selecting a sensor subset. A systematic approach is needed.

4. Projection methods

Principal Component Analysis (PCA) or Linear Discriminant Analysis (LDA) have
been used to map the n-dimensional sensors space onto 2-dimensional feature space.
Over this new space sensor selection can be done using the following rule: features with
minimum loadings are eliminated. PCA use eigenvectors and eigenvalues to define the
subspace orthogonal base from the data covariance, trying to preserve the variance
presented by the original N dimensional distributed data into the subspace.
On the other hand, LDA tries to generate a data projection maximising distance between
a priori classes. LDA is a supervised technique: the number of classes and the label of
each input pattern is given to the algorithm. 2-D planes from PCA and LDA are
presented in figure 2.

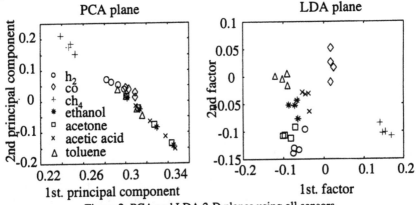

Figure 2: PCA and LDA 2-D planes using all sensors

Figures shows that in the LDA plane clusters appear not as mixed as in the PCA plane.
Sensors are, in fact, a mathematical base of the feature space, and angle between sensor
vectors and the PCA/LDA plane contains information about the sensor relevance. More
orthogonal sensors contain less information than the others, and they can be eliminated
[3]. Process to select sensors can be summarised in three points.
a) Find 2-D PCA subspace and/or 2-D LDA subspace
b) Find the angle between every sensor vector and the PCA subspace and/or LDA
   subspace.
c) Eliminate the sensor with the biggest angle
d) If number of features grater than 1, Go to a) else end.
With PCA and K-NN classifier there is no improvement of the recognition rate when
any subset of sensors is selected.
Using the same strategy with LDA, an improvement is found when the five more
orthogonal sensors are eliminated. Results are presented in figure 3.

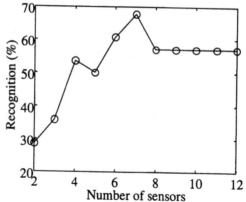

Figure 3. Recognition rate for de LDA strategy.

5. Search methods

In the search methods the maximisation or minimisation of an objective function guides the selection. The objective function used for the problem presented above is the recognition rate (wrapper).

In principle, an exhaustive search of the best result can be done, but the number of combinations to evaluate is 2^N, where N is the number of features. If N is high this exhaustive search is impossible in practice. In the case exposed, we may use exhaustive search as a reference procedure to check the performance of the different algorithms. Among the 4095 possibilities, the best result is a configuration with only 5 sensors that presents a recognition rate of 75 %. If we order the all-possible subsets according the recognition rate, the configuration that uses all sensors is at position 93, far from the best. At this point it must be highlighted that the best recognition rate only reaches 75% and this value has to be taken as reference for the other algorithms. The recognition rate depends strongly with the problem and with the classifier used. In our case the classifier is K-NN, and surely other classifiers would improve the 75% of recognition rate but that is not the objective of this work.

5.1 Sequential Forward and Sequential Backward Selection

Among the different search algorithms, we remark the Classical Sequential Forward Selection (SFS) and the Classical Sequential Backward Selection (SBS) [2].

With SFS the selection algorithm starts from the empty space, and a feature is added if provides the best improvement of the objective function. SBS uses the opposite strategy; the algorithm begins from the whole feature space and a feature is extracted if provides the less descent of the objective function.

Figure 4 presents results for both methods and it can be seen that SFS present better results (64% of recognition rate) than SBS (60% of recognition rate). However, these subsets are not the best ones and even the selection provided by LDA is better. On the other hand, it is logical that these methods do not find the best solution because they only explore a small percentage of the whole set of configurations. This aspect is reflected in Table 1 where the number of configurations evaluated by every method is listed.

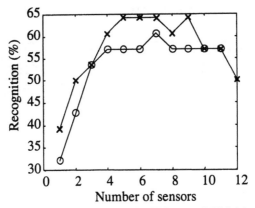

Figure 4. Recognition rate for SFS (x) and SBS (o)

5.2 Genetic Algorithms (GA)

In this section we describe the use of Genetic Algorithms (GA) for feature selection [4][5]. In fact, the problem of selecting features with an objective function is an optimisation problem. The GA is an optimisation technique based on the Darwinian evolution theory: "survival of the fittest". GA are specially suited for this kind of problems, where every feature can be easily codified in the chromosome: a binary string where one means the feature presence and zero means the feature absence.

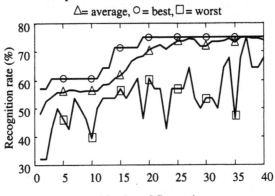

Figure 5. Evolution of the recognition rate for GA method

At every generation, the recognition rate is evaluated for every individual within the population. The fittest individuals are selected for mating using the genetic operators of crossover and mutation. The offspring constitute a new generation.

Parameters used for the evolution of the population were: Population size: 30 chromosomes. Initial population: random. Selection: tournament. Crossover: two point with probability 0.8. Mutation probability: 0.008.

With GA the best configuration and, consequently, the best recognition rate is found. As Table 1 shows, this result is achieved exploring only 600 configurations from the all

4095, less than 15% of combinations. Figure 5 shows the evolution of the recognition rate.

Table 1: Comparison of the best result found by the different methods													
Sensors	1	2	3	4	5	6	7	8	9	10	11	12	Recog. rate	N° of eval.
All sens.	1	1	1	1	1	1	1	1	1	1	1	1	57%	1
LDA	1	0	1	1	0	0	0	1	1	0	1	1	67%	12
SFS/SBS	1	0	1	1	0	1	0	1	1	0	1	0	64%	78
GA	0	0	1	1	0	0	1	0	1	0	0	1	75%	600
Exhaust.	0	0	1	1	0	0	1	0	1	0	0	1	75%	4096

6. Conclusions

Feature selection is a simple method for improving the e-nose performance. It allows the development of more simple classifiers and less data points for tuning are needed.

The feature selection method based on the angles to the PCA plane is not suited for subset selection in pattern recognition if principal components do not carry the discerning information. The equivalent method base on LDA provides better results than PCA, because it finds the optimal projection concerning class separability. It provides good results, even better than SFS and SBS, it is low time consuming but the best solution is not warranted.

SFS and SBS are less computational intensive than GA, and both provide good results but, again, they do not warrant the best solution.

Among the different methods tested, GA provides the best result in terms of final recognition rate, exploring only fifteen percent of the whole possibilities. GA are specially suited to make feature selection, where feature can be easily codified in the chromosome.

Acknowledgements

This work has been partially founded by a Spanish project: CICIT TIC 98-0987-C03-03

References

[1] R.O.Duda and P.E. Hart. Pattern Classification and scene Analysis. *New York: Wiley, 1973*

[2] S. Stearns, On selecting features for pattern classifiers. *3-d International conference on Pattern Recognition, Coronado, CA, pp 71-75 (1976)*

[3] A.N. Chaudry, T.M. Hawkins and P.J. Travers. A method for selecting an optimum sensor array. *Proceedings on ISOEN 99, Tübingen, September 99 pp 179-182 (1999)*

[4] B. G. Kermani, S. S. Schiffman and H. T. Nagle. A novel method for reducing the Dimensionality in a Sensor Array. *IEEE Transactions on Instrumentation and Measurement Vol 47, N° 3 pp 728-741 (1998)*

[5] P. Corcoran, J. Anglesea, M. Elshaw. The application of genetic algorithms to sensor parameter selection for multisensor array configuration. *Sensors and Actuators A, vol 75 pp 57-66 (1999)*

Section 3: Data Processing
Paper presented at the Seventh International Symposium on Olfaction and Electronic Noses, July 2000

89

Odour Identification under Drift Effect

Cosimo Distante[1], Tom Artursson[2], Pietro Siciliano[3], Martin Holmberg[2], Ingemar Lundström[2]

[1]Dipartimento di Ingegneria dell'Innovazione, Università di Lecce, via Arnesano – 73100 Lecce, Italy, Phone: +39 0832 320253, Fax: +39 0832 320341, E-mail: cosimo.distante@unile.it

[2]S-SENCE and Laboratory of Applied Physics, Linköping University, S-581 83, Linköping, Sweden, Phone: +46 13 288904, Fax: +46 13 288969, E-mail: tomar@ifm.liu.se

[3]Istituto per lo Studio di Nuovi Materiali per l'Elettronica (IME) – C.N.R. via Monteroni – 73100, Lecce, Italy
Phone: +39 0832 320244, Fax: +39 0832 3253495, E-mail: pietro.siciliano@ime.le.cnr.it

Abstract. This paper presents a solution to the parameter drift problem, by using a neural network based on multiple Self-Organizing Maps. mSOM. This architecture has been tested on two complex data sets where the results are successful. The network is able to adapt itself to new changes of the input probability distribution by repetitive self-training processes based on its experience.

1. Introduction

Gas sensors transduce chemical properties to electrical signals. The chemical properties are the concentration of different species, atoms or molecules. Since many sensors, multisensing, are often used to characterize a chemical environment, pattern recognition methods are used to extract information from the distributed response. The combination of gas sensors and multivariate algorithms is capable to detect and recognize complex odors for e.g. quality assurance of food [1,2], paper [3] and grain [4] or for monitoring pharmaceutical processes [5] and environmental aspects such as car exhaust gases [6].
The response of chemical sensors is dependent of several other factors than the chemical environment that is under study.

Drift is a dynamical process, caused by physical changes in the sensors and the chemical background, which gives an unstable signal over the time. The drift could be both reversible, e.g. condensation of vapor on the sensors, and irreversible, aging. The samples and the operator, via contamination of the instrument, could introduce drift. One kind of drift caused by the measurement history is memory effects. This means that measurement at time t is highly influenced of measurements at time $t-n$. This leads to that the same gas mixture will not give one well defined pattern. The drift causes the pattern recognition models to be very short-lived. If no drift correction of the sensor signals is made, the models will have a continuous need for recalibration. Since the training phase of the pattern recognition model should contain all future coming variance a lot of samples are necessary. In real applications, processes, the samples may be very expensive which makes it impossible to recalibrate the pattern recognition model often. Basically, there are two ways to obtain long-term reproducible and reliable gas-sensor array based instruments and both of them are necessary. The first, and most obvious, alternative is to improve the sensors to give a more stable long-term performance. The second approach is to refine the calibration models so the influence of drift will be reduced.

Different approaches to correct gas sensor data that suffer from drift have been tried. One way to monitor the instrument performance over time is to measure calibration standards, reference gas. Attempts have been made using a reference gas as a reference value and then correcting all subsequent readings accordingly.[7, 8] Component correction [9] is new interesting method using PCA and PLS algorithms to find the drift direction from measurements of a reference gas. This direction is then removed from the samples. Fundamental studies of the mathematical properties of the drift effects have been studied by Davide et al.[10] It was seen that the behavior was different in different frequency ranges. This group has also developed successful pattern recognition models based on system identification theories.[11, 12] These models do not require the use of reference gas. Adaptive resonance theory has also been used as recognition of measurements subjected to drift [13].

This paper introduces a new drift counteraction method based on many parallel adaptive self-organizing maps, called mSOM.

2. mSOM

Several works make use of the basic SOM [14] for classification of chemical patterns. Then the labeling phase follows the unsupervised one, in order to classify data based on the Euclidean distance. However, in the context of electronic nose, the use of a single map often becomes useless due to drift. If a neuron is not often (or never) activated, it would not map the new probability distribution of the odor. In other words, if a cluster moves to a new position, it is not obvious that all the neurons belonging to that cluster will be updated. This behavior could give rise to confusions, since in the middle of a cluster there could be a neuron that belong to another cluster and it has not been activated

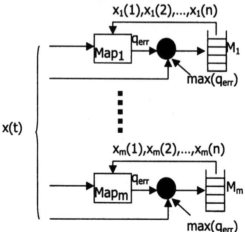

Figura 1 mSOM architecture.

since very long time. For this reason we developed a multi self-organized map that assures the self-adjusting process to all the neurons in the local map, and autonomous adaptation to new situations. Our approach preserves the self-organization paradigm by considering as many maps as the various odors are taken under consideration, in order to accomplish the classification task. There have been published other papers using multiple self-organizing maps for classification only, without self-retraining mechanisms [15]. The novelty in this paper is the adjustments of the individual maps in time to be able to predict the gas measurements which suffered from drift. Mainly the processing is made in two phases: training and testing with self-training. The last part is not driven by a normal teacher but is completely autonomous and is based on the past history of each map that is contained into local memories. The quantization error controls the retraining process. Each map self-organizes its codebooks and then refines them by using a learning vector quantization algorithm, LVQ [14], in order to reduce the high uncertainty accumulated at the borders of two or more different clusters (in this case the boundaries of the maps).

In the initial training phase, each map is trained with the associate odour sample, i.e. a set of preprocessed[1] data vectors x. In figure 1, the mSOM architecture is shown. After training, the maximum quantization error is found, based on the data contained in the local memories M_i or in the initial training set. Once the mSOM is trained, the autonomous classification is carried out. At a certain time step t, a preprocessed sensor response x is presented to the network. The distance measure is computed over all the maps, and the winner map (the one with the lowest quantization error) is considered as the estimated recognized odor.

The quantization error is defined with the Euclidean distance and is computed as the average of the distance between all data vectors and the neurons (*codebooks*) of the associated map. Each recognised data vector is stored in the memory of the associated map. After a certain number of iteration, each map is requested to retrain its recognition code by using its local memory M_i. This because the drift changes the sensor response in which at a certain point mSOM is not anymore able to recognize the data and a retraining process is needed. Also, misclassification is mainly caused by noisy data and storing of erroneous data in the local memories, ending up in unstable behavior of the network. To prevent this problem, the maximum quantization error is computed at the end of each training process. In other words, a data vector is stored only if its position is not at the boundaries of the map where high uncertainty is present. Once the maps have been organized in order to approach the input probability distribution, the LVQ algorithm is carried out. LVQ is a useful method to refine clusters, and to reduce the area in between two or more clusters in which highest uncertainty is present. This area is also known as Bayesian borders.

3. Experiment

Two data sets which suffer from drift were used to test the mSOM algorithm. Data set A, used in [9], contained gas mixtures of hydrogen, ammonia, ethanol and ethene, for the

concentrations see table 1. Normally, gas sensors are always annealed before use to remove a strong initial drift. In order to get significant drift effects in a relatively short

---

[1] The preprocessing is carried out by extracting the first useful principal components over the training set.

period of time, the sensors used here in data set A were new and not annealed. More than two months of randomly repeated measurements gave a large data set, 102 measurements on

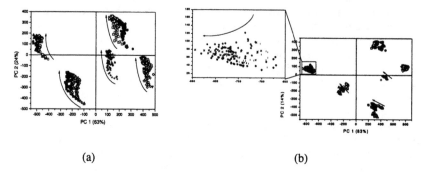

(a)   (b)

Figura 2 Two score plots over the different the data set A and B. The symbols in the figure are the mixtures from the table 1 and cardboard papers: ▲=1, =2, ■=3, ▼=4, o=5, ●=6, △=7, ◊=8, +=9, x=p1, ⊕=p2, | =p3, ◆=p4. The arrows indicate the drift direction.

each of the nine gas mixtures. The data set B contained measurements on the same gas mixtures and concentrations as in data set A, plus cardboard paper measurements. The sample systems used for data set B differed from A, since this one contained glass vessels with cardboard paper. The headspace was sampled by pressing technical air through the glass vessels containing the different paper samples. These measurements were repeated during one month, giving a data set with 13 different gas mixtures each containing 37 measurements. The measurements in data set B suffered from drift but not as much in A since these sensors were not new. For the paper measurements there existed two kinds of drift in parallel, the sensor drift and the chemical drift due to the changes in the paper. The chemical drift is of course interesting to follow. Furthermore, the measurements in data set B contained also memory effects.

The gas sensor instrument contained two arrays of MOSFET (Metal-Oxide-Semiconductor-Field-Effect-Transistor) sensors [16] each containing 10 sensors and two arrays of MOS (Metal Oxide Semiconductor) sensors [17] containing 10 and 9 sensors, respectively, making a total of 39 sensors. These sensors are commercially available. The gas sensors were exposed to 2-min pulses of the test gas in synthetic air. The time between the pulses was 18 min, where the sensors were exposed to synthetic air in order to recover the baseline value. The oxygen level was constant at 20% of the total flow.

The size of each map was 9 by 9 neurones, which were randomly initialised. The first 36 measurements were used for training in data set A and the first 15 in data set B. The rest of the measurements were used as test sets. As preprocessing, the first 15 principal components were calculated on the training data and the test set was then projected down to these PCA models. The principal components were then used as inputs to the mSOM. The learning rate parameter $\alpha=0.5$ which was exponentially decreased and, the refresh rate, which control the retraining process was set to 50 unknown input presentations.

4. Results

An overview of the two data sets is shown in two score plots, see Fig. 2. In figure 2a the pronounced drift in data set A appears clear. The arrows in Fig. 2a and 2b indicate the time order. In data set B the drift is not as strong as in A, see Fig. 2b. For some gas mixtures the drift direction was not that legible, i.e. cluster 2 & 6 and 4 & 8. The memory effects are obvious for cluster 2, 5, 3, 7 and the paper measurements. From the zoom of the paper measurements it is seen that the papers are overlapping each other in the two-dimensional score plot, especially the first measurements in each cluster.

The results of the mSOM classifications for data set A showed only 1.5% misclassifications for the test set, and it is mainly due to mixture 2 which is classified as mixture 6. For the data set B the number of misclassifications is a bit higher, 5.4%, and is among the cardboard paper 1, 3 and 4 but also gas mixture no. 4 is classified as gas

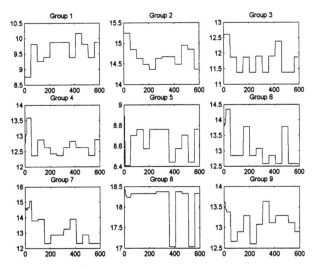

Figure 3. Quantization error for the data set A.

mixture no.8 four times. To test the network, the objective function to be reduced is the average quantization error recorded over several retraining processes and is defined as follows:

$$qerr_{ave} = mean_{k \in M_j} \sum_{i \in Map_j} \|x - m_i\|_2$$

where M_j is the j-th memory, x the data vector and m_i the codebook of the j-th map. In general, when a cluster moves away from its original position because of the drift, the resulting average quantization error computed for that cluster will increase. Our objective is to keep the average quantization error as low as possible, in order to have the map always on top the data vectors belonging to that cluster.

Figure 3 shows the quantization error graph related to each map, and it is interesting to note that the picks are related to the retraining process that move the neurones toward the new input probability distribution, thus reducing the dinstance between the new cluster location and the corresponding map neurons.

Table 1 The concentration (ppm) of the individual gases in the nine different gas mixtures.

Group no.	H2	NH3	C2H5OH	C2H4
1	10	10	20	80
2	80	10	20	20
3	10	80	20	20
4	80	80	20	80
5	10	10	80	20
6	80	10	80	80
7	10	80	80	80
8	80	80	80	20
9	45	45	50	50

5. Discussion

The new approach we have proposed, mSOM, presents several attractive features and proves to be a tool for avoiding continuous recalibration in a sensor system. For data set A, which contained nine gas mixtures with a pronounced drift in all the clusters the misclassification was very low, only 1.5%. This is only 9 errors out of 594 measurements. For the other data set, B, which besides the nine gas mixtures also contained cardboard paper measurements, the mSOM managed to classify 94.6% right. Even though most of the misclassifications are among the paper measurements, mSOM manage to reduce the influence of drift and separate the overlapped clusters, see Fig. 2b.

mSOM presented in this paper is a powerful tool for pattern recognition of measurements subjected to drift. This depends on the fact that it is an adaptive method which follows the gas mixtures very well in time. An unknown sample presented to the network is classified according to the best-matched map and is then used to update the weights of that map. In the case of the cardboard paper experiment would have been interesting to follow how the different papers changes in time. But since mSOM follows both the drift from the sensors and the chemical drift caused by the aging of the paper it is impossible.

6. Acknowledgement

Three of the authors research on chemical sensors is supported by grants from the Swedish Research Council for Engineering Science and the National Swedish Board for Industrial and Technical Development (NUTEK). Swedish Industry, NUTEK and Linköping University finance S-SENCE (Swedish Sensor Centre). One of the authors would like to thank the NOSE Network on Artificial Olfactory Sensing for giving him the opportunity to visit the group of Pietro Siciliano at University of Lecce. The cardboard paper was received from Gunnar Forsgren, Stora Enso.

7. References

[1].   Eklöv T, Johansson G, Winquist F, Lundström I. Monitoring sausage fermentation using an electronic nose. *J. Sci. Agric.* 1998; **76**:525-532.

[2].   Siciliano P, Rella R, Quaranta F, Epifani M, Capone S, Vasanelli L, Analysis of vapours and foods by means of an electronic nose based on a sol-gel metal oxide sensor s array, Sens. and Actuators B in press.

[3].   Forsgren G, Frisell H, Ericsson B. Taint and odour related quality monitoring of two food packaging board products using chromatography, gas sensors and sensory analysis. *Nord. pulp paper res.* 1999; 14:4-15.

[4].   Börjesson T, Eklöv T, Jonsson A, Sundgren H, Schnürer J. Electronic nose for odor classification of grains. *Cereal Chem.* 1996; 73:457-461.

[5].   Bachinger T, Mårtensson P, Mandenius C-F. Estimation of biomass and specific growth rate in a recombinant Escherichia coli batch cultivation process using a chemical multisensor array. *J. Biotechnol* 1998; 60:55-66.

[6].   Tobias P, Mårtensson P, Baranzahi A, Salomonsson P, Lundström I, Åbom L, Lloyd-Spetz A. Response of metal-insulator-silicon carbide sensors to different components in exhaust gases. *Sens. Actuators B* 1998; 47:125-130.

[7].   Fryder M, Holmberg M, Winquist F, Lundström I. *A calibration technique for an electronic nose,* Proceedings from Transducers '95 and Eurosensors IX, Stockholm, Sweden, 1995, 683-686.

[8].   Haugen J E, Tomic O, Kvaal K. A calibration method for handling the temporal drift of solid state gas-sensors. *Anal. Chim. Acta* 2000; 407:23-39.

[9].   Artursson T, Eklöv T, Lundström I, Mårtensson P, Sjöström M, Holmberg M. Drift correction for gas sensors using multivariate methods. *J. Chemometrics* 2000; 14: 711-724.

[10]. Davide F, Di Natale C, Holmberg M, Winquist F. *Frequency analysis of drift in chemical sensors,* Proceedings from 1 st Italian Conference on Sensors and Microsystems, Rome, Italy, 1996, 150-154.

[11]. Holmberg M, Winquist F, Lundström I, Davide F, Di Natale C, D'Amico A. Drift counteraction for an electronic nose. *Sens. Actuators B* 1996; 35-36:528-538.

[12]. Holmberg M, Davide F, Di Natale C, D'Amico A, Winquist F, Lundström I. Drift counteraction in odour recognition application: lifelong calibration method. *Sens. Actuators B* 1997; 42:185-194.

[13]. Distante C, Siciliano P, Vasanelli L, 2000, Odor discrimination using adaptive resounance theory. *Sens. and Actuators B* 3461.

[14]. Kohonen T, 1997 Self-Organizing Maps, Springer.

[15]. Cervera E, del Pobil A P,1996, Multiple Self-Organising Maps: a hybrid learning scheme, Neurocomputing vol. 16, pp. 309-318.

[16]. Lundström I, Spetz A, Winquist F, Ackelid U, Sundgren H. Catalytic metals and fiel-effect devices - a useful combination. *Sens. Actuators B* 1990; 1:15-20.

[17]. Göpel W,Schierbaum K D. SnO_2 sensors: current status and future prospects. *Sens. Actuators B* 1995; 26-27:1-12.

Section 3: Data Processing
Paper presented at the Seventh International Symposium on Olfaction and Electronic Noses, July 2000

97

Electronic Noses using "Intelligent" Processing Techniques

A. Cremoncini[*], F. Di Francesco[**], B. Lazzerini[+], F. Marcelloni[+], T. Martin[++], S.A. McCoy[++], L. Sensi[*], G. Tselentis[+++]

[*]I.S.E. Ingegneria dei Sistemi Elettronici s.r.l. – Vecchiano (PISA) – ITALY
Phone: +39.050.804.343, Fax: +39.050.804.727, E-mail: mbxise@tin.it
[**]Centro Piaggio – University of Pisa – Pisa (Italy)
[+]Dipartimento di Ingegneria della Informazione – University of Pisa
[++]Department of Engineering Mathematics – University of Bristol (UK)
[+++]MIT Management Intelligenter Technologien GmbH – Aachen (DE)

Abstract. This paper presents results of ESPRIT 25254 (INTESA) R&D project. The Project Involved two Industrial Enterprises and three University Departments from three European Countries over a 2,5 year period, and had the aim to develop and validate an innovative approach for processing Chemical Sensor Data. A number of Prototypes of Artificial Nose devices were manufactured, and many industrial sectors of application were investigated throughout the Project.

1. Introduction

The INTESA Project activity was directed at investigating the use of "Uncertainty Modelling" techniques on Chemical Sensor data, with a twofold aim: first, to understand more about "where" and "how" information is embedded in chemical sensor response curves and, second, to work out a solution of a Processing Software package for "Electronic Noses" with improved performance for perspective application to problems of real "Industrial" Interest.
Several prototype configurations were implemented, using both MOS and CP sensors.

2. Overall description of the INTESA Philosophy

The basic problem addressed by the INTESA project is the development of software processing techniques and tools for information extraction from chemical sensors data, with the ultimate scope of obtaining a degree of knowledge similar, in some way, to the one obtained by a human operator.

In the classical analysis tools (gas chromatography or mass spectrography) a gas sample is broken down into its individual components, but the INTESA prototype takes an odour "fingerprint" trying to reproduce a human judgment of the odour sample. From a general point of view, we have to keep in mind that biological olfaction is certainly superior to machine smelling, and the actual chemical sensors are limited in selectivity and sensitivity performances; so we can be satisfied in reaching good results in some particular field of industrial interest.

The INTESA Prototype system (see Fig. 1) is composed of two main parts: the chemical sensor array (from Centro E. Piaggio, University of Pisa) with the related hardware interface, and the software modules for control, data processing and archiving. The software components reside on a high-end personal computer which also hosts the front-end electronics for sensor data acquisition.

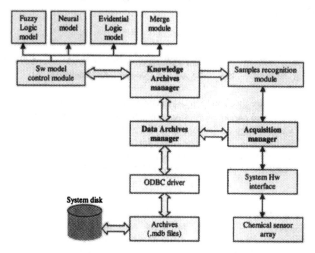

Figure 1: INTESA Prototype System block diagram

The system integrates the following processing models (the resulting integration follows a scheme developed and implemented by ISE s.r.l. with the Consortium partners collaboration):

- Fuzzy logic model: constitutes the fuzzy logic classifier of the system, and has been developed by the Department of Information Engineering of the University of Pisa.
- Evidential logic model: constitutes the classifier based on evidential logic theory, and has been developed by the Department of Engineering Mathematics of the University of Bristol.
- Neural network model: constitutes the neural network based classifier, and it has been developed by MIT GmbH.
- Merge, or "Integrating" module: each of the three classifiers processes input data, and it produces an output as an intermediate result; eventually, three different classifier outputs are "merged" together by means of a statistical analysis, which provides the final prototype output.

Processing models were validated using a K-fold strategy, were each data set of N elements was divided in K continuous in time groups (folds) of the same length, and then each fold was used as the test set for the classification models which had been trained with the remaining data. Given the fact that the purpose of the Project was to improve performance classification vs. "conventional" processing techniques, one of the objectives of the experimental activity was to try to produce large consisting datasets, acquired over significant periods of time (where conventional methods tend to give poor results, due to "sensor drift" phenomena). In the following, the three processing models will be described in more detail, and an outline of most important project results will be given.

3. "Intelligent" Processing Software Models

3.1. *"Fuzzy Logic" Model*

We adopted a method based on fuzzy logic to model the general shape of typical responses of sensors to odours (Lazzerini, 1998). The method makes use of fuzzy linguistic expressions, and exploits these expressions to recognise unknown responses.

Let us consider a signal, i.e., a single response of sensor s, $s=1..Q$, to odorant o, $o=1..O$. With reference to Fig. 2, the time space (i.e, the time interval during which samples are taken) and the signal space (i.e., the real interval to which the amplitude values of the signal belong) are uniformly partitioned into H and K sub-intervals (possibly, $H \neq K$). In both spaces, a triangular fuzzy set is built for each sub-interval extreme, which is the fuzzy set modal value (i.e., the point at which the membership degree is 1). Each fuzzy set covers two adjacent sub-intervals, with the exception of the first and last fuzzy sets that cover one sub-interval. A label is associated with each fuzzy set, thus producing a linguistic partition of the two spaces. For each signal, the size of the signal space is chosen to be equal to the dynamic range of the signal itself. This produces a normalisation of the signals so that they can be analysed independently of their amplitude.

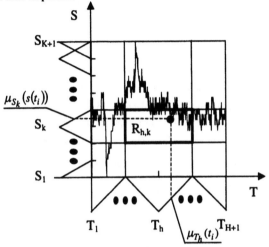

Figure 2: Fuzzy partition of the time and signal spaces

Figure 2 shows how fuzzy sets on the time space are associated with fuzzy sets on the signal space. Given two fuzzy sets T_h and S_k, the Cartesian product of their supports defines a region $R_{h,k}$ in the plane. The fuzzy set S_k is said to be *active* when the input is T_h, if $R_{h,k}$ contains at least one signal sample. For each fuzzy set T_h, $h=1..H+1$, we associate an *activation value* $a_{h,k}$ with each fuzzy set S_k, $k=1..K+1$, as follows:

$$a_{h,k} = \frac{\sum_{i=1}^{P} \mu_{T_h}(t_i) \cdot \mu_{S_k}(s(t_i))}{\sum_{i=1}^{P} \mu_{T_h}(t_i)}, \tag{1}$$

where P is the number of signal samples contained in $R_{h,k}$, t_i is the ith time sample, $s(t_i)$ is the signal value at t_i, μ_{T_h} and μ_{S_k} are the triangular membership functions of T_h and S_k, respectively. The rationale for this computation is that a sample close to the modal value of a fuzzy set on the time or signal space should contribute to the activation more than samples with a low membership degree. For each fuzzy set T_h, we define a fuzzy set $Y_h = \left\{ a_{h,1} \middle/ S_1 + ... + a_{h,K+1} \middle/ S_{K+1} \right\}$ in the space of the labels S_k, the activation value $a_{h,k}$ being the membership degree $\mu_{Y_h}(S_k)$ of S_k to Y_h. The linguistic representation of the response of sensor s to odorant o is the classical set $L_{s,o} = \{Y_1,...,Y_{H+1}\}$.

For each pair sensor/odorant (s,o), $s=1..Q$, $o=1..O$, we build a linguistic fuzzy model called *sensor-odorant linguistic model*. The sensor-odorant model is obtained by considering a set of responses of the sensor s to the odorant o in repeated experiments (*training set*). For each response in the training set, a linguistic representation is generated. Choosing the size of the signal space equal to the dynamic range of each signal allows us to make the responses in the training set comparable with each other. The time and signal spaces of all these responses are divided, respectively, into the same numbers H and K of sub-intervals. The optimal numbers of the fuzzy set modal values in the time and signal spaces are application-dependent parameters. While corresponding sub-intervals of the time space in different responses have the same size, the size of corresponding sub-intervals in different signal spaces is proportional to the size of the pertinent signal space. Let us assume that the training set consists of N responses. Then the *sensor-odorant linguistic model* that describes linguistically the pair (s,o) is $M_{s,o} = \left\{ \overline{Y}_1^{s,o}, \dots, \overline{Y}_{H+1}^{s,o} \right\}$, where, for $h=1..H+1$,

$$\overline{Y}_h^{s,o} = \left\{ \overline{a}_{h,1}^{s,o} \Big/ S_1 + \dots + \overline{a}_{h,K+1}^{s,o} \Big/ S_{K+1} \right\}, \quad \overline{a}_{h,k}^{s,o} = \mu_{\overline{Y}_h^{s,o}}(S_k) = \frac{1}{N} \sum_{n=1}^{N} a_{h,k,n}^{s,o} \quad \text{and} \quad a_{h,k,n}^{s,o} \text{ is the}$$

activation value associated with the label S_k in the nth response of the training set. This means that, for each fuzzy set T_h, the activation value of the fuzzy set S_k is calculated considering the average of all the signal samples belonging to the rectangles $R_{h,k,n}^{s,o}$, $n=1..N$.

To classify an unknown odorant, the signal produced by the sth sensor is converted into a linguistic fuzzy model $L_{s,u}$ using the same input/output partitions as the models $M_{s,o}$. Then, $L_{s,u}$ is compared with each $M_{s,o}$. The comparison produces the *shape match value* $SM^{s,o}$ between $L_{s,u}$ and $M_{s,o}$, defined as the following weighted sum:

$$SM^{s,o} = 1 - \sum_{h=1}^{H+1} w_h^{s,o} \cdot \sum_{k=1}^{K+1} \left| \mu_{Y_h^{s,u}}(S_k) - \mu_{\overline{Y}_h^{s,o}}(S_k) \right|, \tag{2}$$

where $w_h^{s,o}$ are real numbers such that $\sum_{h=1}^{H+1} w_h^{s,o} = 1$. The weights $w_h^{s,o}$ are chosen so as to give more importance to the fuzzy sets $\overline{Y}_h^{s,o}$ which better characterise the pair (s,o).

Given the N responses in the training set, $w_h^{s,o}$ is computed as $w_h^{s,o} = \dfrac{v_h^{s,o}}{\sum\limits_{h=1}^{H+1} v_h^{s,o}}$, where

$$v_h^{s,o} = \sum_{n=1}^{N} v_{h,n}^{s,o}, \quad v_{h,n}^{s,o} = 1 - \sum_{k=1}^{K+1} \frac{1}{K+1} \left| \mu_{Y_{h,n}^{s,o}}(S_k) - \mu_{\overline{Y}_h^{s,o}}(S_k) \right| \quad \text{and} \quad \mu_{Y_{h,n}^{s,o}}(S_k) \text{ is the}$$

membership degree of S_k to the fuzzy set $Y_{h,n}^{s,o}$ of the nth response relative to odorant o and sensor s in the training set.

For each odorant o, the *odorant shape match value* is obtained as $SM^o = \sum\limits_{s=1}^{Q} w^{s,o} \cdot SM^{s,o}$,

where $w^{s,o}$ are real numbers such that $\sum\limits_{s=1}^{Q+1} w^{s,o} = 1$. The weight $w^{s,o}$ takes into account the confidence of the sensor s in recognising the odorant o. In general, the closer the signals generated by a sensor in repeated experiments with the same odorant, the better the sensor with respect to that odorant. The weight $w^{s,o}$ is inversely proportional to the average width of the linguistic fuzzy sets in $M^{s,o}$. The width of a linguistic fuzzy set is the number of labels in its support. The rationale behind this design choice is that narrow models correspond to more reliable sensors.

When an unknown odorant is presented to the system, the pattern recognition system recognises it as the odorant with the highest shape match value.

3.2. *"Evidential Logic" Classifier*

The classifier from the University of Bristol is based on processing the transient response produced by a sensor when it is exposed to odour.

The classifier is an extension of the data browser approach (Baldwin and Martin, 1992, 1995, 1997), whose aim is to replace large volumes of tabular data with compact summaries of the underlying relationships in the form of Fril rules. Fril (Baldwin, Martin and Pilsworth 1988) is a logic programming language extended with facilities to handle uncertainty. In general, the price paid for summarising data in a concise form is a loss of precision - the rules are uncertain in some aspects. However, as shown by fuzzy control applications, this loss of precision can lead to considerable strength and robustness and additional advantage arise because rule-based summaries are generally easy for a human to understand.

The data browser can generate standard Fril rules or Fril evidential logic rules, in which the features are weighted by importances. The importances can be found automatically by semantic discrimination analysis. The data browser can be used for incremental learning, in which the rules are updated each time new training data become available (Baldwin, Martin and McCoy 1998). As this facility was not available in other classifiers, it has not been used in this study.

Since the amount of training data is limited, fuzzy sets are used to generalise the training data when forming rules. The measured values are transformed into fuzzy sets using mass assignment theory (Baldwin, 1992),(Baldwin, Martin and Pilsworth 1995), which gives a theoretical framework for handling statistical and fuzzy uncertainty. The method thus has a sound underlying mathematical basis unlike the ad hoc methods used frequently in fuzzy control approaches.

Observing several sensor responses it can be seen that, even if the response amplitude differs from experiment to experiment, the curves present similar shape. Therefore, the shape of the response is the information that must be abstracted into the fuzzy model.

Shape similarity can be easily recognised by human beings, and can be described linguistically. For example, a possible description of a sensor response is: "... almost constant during the stabilisation phase; followed by a fairly rapidly increase during the in phase and a slow decreases during the out phase". The linguistic description is not affected by minor changes in the sensor response, such as different amplitude or noise. This information can be expressed by means of gradient values; but this mathematical model does not have the same abstraction capacity as the linguistic description.

A higher abstraction level can be reached if fuzzy gradient values are used to represent curves. In this model, the key words become labels of several fuzzy sets whose membership functions are automatically learned from examples. The number of fuzzy gradient values necessary to represent a response depends on the shape of the curve. A new gradient value should be taken when a major change in the gradient occurs. At the present stage of the work, the number and the position of the interval on which the gradient values are calculated are design parameters of the classification system. They are established by the designer by visual inspection of the sensor responses. The automation of this decision is part of the future development of the classification method.

There are two levels of rules in the classifier. In stage 1, rules are derived for each sensor/substance of the form

> Sensor <i> indicates substance <j>
> > IF   in phase gradient is *Steep-Increase*
> > AND   out phase gradient is *Slow-Decrease*

where italicised terms are fuzzy sets generated from data. In the work quoted here, four to six gradient terms were used to describe each response curve. In stage 2, a further set of rules fuses the results from each sensor:

Classification is substance <j>
      IF *most* sensor <i1> indicates substance <j> (importance w1)
      AND   sensor <i2> indicates substance <j> (importance w2)
         ...
      AND   sensor <im> indicates substance <j> (importance wm)

This system can be used to evaluate the support for a given test example being a member of each of the defined odour classes. The class with the largest support is the predicted output class produced by the classifier.

In some cases, questions can be raised over the assumption that a single distribution of feature values is sufficient to characterise the response of a sensor to an odour class. In such cases, points belonging to the same class may appear as multiple clumps of data points, or thinly scattered across a large area, overlapping the distributions for other classes considerably. The underlying reasons for such widely spread points in the dataset may include poor control of the measurement experiments, drift in the responses of sensors over time, or inappropriate features which do not provide useful discrimination between classes. A refinement of the method uses multiple clusters of points for each odour class to model the response of each sensor. Each cluster is associated with a number of fuzzy sets, one for each feature, modelling the typical values of those features for points in that cluster. This requires a pre-processing step, details of which appear elsewhere.

This method was developed for classification using conducting polymer sensors, where the shape of the response can be related to the underlying physico-chemical changes. In metal oxide sensors, the response is much faster and it is not clear that the details of the response curve are so relevant. The base resistance and overall change in resistance were included as additional features, with evidential logic rules used in stage 1 so that importance could be estimated.

The weight assigned to each of the sensors in stage 2 rule is calculated from the success rate of the corresponding stage1 rule in correctly classifying the examples in the training set belonging to the appropriate class. This method also enables an automatic ranking of importance of sensors in the classifier. As an additional experiment, the data from the most important sensor (according to this method) was used to generate a decision tree using the mass assignment ID3 method (Baldwin, Lawry and Martin 1997).

3.3. *"Neural Network" Classifier*

The procedure used by the Neural Network to classify raw data can be depicted as follows:

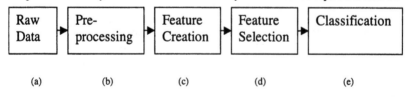

    (a)   (b)   (c)   (d)   (e)

Figure 3: Schematics of Neural Model Classification Procedure

By Raw Data we mean the sensor output, by Pre-processing any method that smoothes the signal trying to de-noise it, by Feature Creation any (for the moment arbitrary) simple characteristics of the pre-processed data (like slopes, peaks, max, min etc.), by Feature Selection the optimal selection of these features according to the target classes and by Classification the creation of a model that can answer the initial classification problem. In steps (a)-(e) there is an attempt to decrease dimension and at the same time increase discriminative information.

Pre-Processing

The first step after gathering information from the sensors is pre-processing where a normalisation and a de-noising filtering is performed. The normalisation and the reset of the signals is performed for two reasons: first, for using a common base of comparison for the signals regardless initial settings of the control variables (like the electrical current that flows in the sensors) and second for providing at a later stage a balanced input for the NN classifier. The general filtering procedure that was selected (algorithm based on the Savitzky-Golay method, 1964) is a smoothing filter as it was considered that the trend of the signal was decisive for evaluation and that high frequency components were due to electrical noise.

Feature Creation

In this stage we decide which are the potential discriminative features from the pre-processed signals or with simple transformations we create new ones. This procedure is somehow an arbitrary one but some indications are given by simple visual inspection or statistical analysis of the raw data. Visual inspection of the time signals or information on the acquisition protocol could not guide us during this selection. The idea is to generate as much as possible potential discriminative features and let the following stage of feature selection to pick the more important ones. So, we considered statistical values from the reset signal   and we produced also a signal of simple first differences :

, $t=1...n$ and where n is the total of data points available.

We calculated for the total length of both signals the following features:

(1)   min(all signal)
(2)   index of min(all signal) which is the time t that the min value appeared,
(3)   max(all signal)
(4)   index of max(all signal) which is the time t that the max value appeared
(5)   max-min(all signal)
where all signal={ , }

The two signals   and   are also segmented k areas (in our application k=12 resulting in areas of 10sec). For these areas the following statistical values are calculated:

(1)   min(area),
(2)   max(area),
(3)   mean(area),
(4)   std(area),
(5)   max-min(area),
where area has k length.

Feature Selection

Feature Selector (FS) is actually a distance classifier that performs classification tests with possible feature combinations. The problem is that the number of testing all possible combinations increases in a factorial function of the size of the feature vector. To tackle this computational burden the algorithm of FS starts measuring the performance of one feature, then proceeds with couples, triplets, etc. but every time it performs a reduced number of evaluations as it takes under consideration how the features performed in the previous algorithm loop. For this reason the algorithm uses two criteria: reclassification error and class discrimination in terms of distance between the classes. The lower the reclassification error and the higher the distance, the better the feature perform to distinguish the classes.

At the end, a feature vector of pre-determined length that classified better the target classes in terms of the two aforementioned criteria is presented at the FS output. For more on FS see Strackeljan et. al. (1997).

Classification Using a Multi-Layer Perceptron MLP

The Neural Network approximates every non-linear mapping of the form $y = f(x)$. Once the feature selection step has been completed, the selected feature combinations are used for classifier design. Test runs ascertain whether the learning phase was successful or not. The classification error for this test data set is usually taken as a measure for the generalisation ability of the neural network.

Several Multi-Layer Perceptron (MLP) schemes were tested but the results revealed that the important issue is rather the choice of the appropriate features rather than the choice of an optimum architecture. Nevertheless, a simple automated procedure for the optimisation of the architecture of the MLP was used: the ten most popular architectures during the analysis phase are stored in the module and a quick run selects the one with the smallest training error. This optimisation should be automated in order that an inexperienced user could benefit from it without having to tune the MLP. What is also important is to keep small the number of the input features specially if we want to have credible training of the network in case of a small knowledge base. So the previous stages of Feature Creation and Feature Selection are critical for the classification.

For the majority of tests an MLP with a single hidden layer and an hyperbolic tangent transfer function and a back-propagation learning method was chosen. The number of input (features) variables was attempted to be kept as small as possible due to lack of sufficient number of examples. The performance of the classifier has to be measured by setting different pairs of training-testing sets and it is mainly referred to the classification results of the testing set.

4. Experimental Results

The above data processing philosophy was evaluated using two prototypes of purposely developed system configurations, on two different applications: waste waters disposal (using a system with 8 conducting polymer sensors), and off-odour evaluation in food packaging quality control (using a system with 6 commercial MOS thick-film sensors).

Waste water samples were collected from three basins at a waste water treatment plant, one containing untreated industrial waste waters, another containing partially treated industrial waste waters and the last containing untreated municipal waste waters. Some samples were diluted in order to lower the headspace substance concentration; eventually, samples were to be classified into nine classes (three different waste basins, in three different dilution each), with samples taken at one week time delay.

The testing was performed using an n-fold cross-validation technique in which the patterns composing the training and test sets were rotated so that in each fold we had a unique training and test set. The average success rate and the standard deviation in the training and test sets were 0.93 and 0.03, and 0.89 and 0.09, respectively.

For the off-odour evaluation in packaging material, the protocol used to acquire data consisted of three phases: i) the stabilisation phase, during which the sensor resistance reached a value considered as base line; ii) the input phase, during which the sensors were exposed to the saturated vapour of the odorant under test; and iii) the output phase, during which the sensors were removed from the odorant to allow the desorption of the odorant under test.

Sampling frequency was 2 Hz. The number of samples in the three phases was 10, 5 and 105, respectively. To simulate the real situation, packages were doped with four different solvents, which are the main cause of bad odours, in 3 different concentrations. The classes of packages doped with the lowest, medium and highest concentration are denoted J, K and Z, respectively. A human panel assessed the packages belonging to classes J, K and Z as good, good, and bad, respectively. In the following, classes J and K will be referred to as J&K. The two different levels of good packages allow us to test the reliability of the electronic nose in recognising slightly different packages belonging to the same class. For each workday, 48 measurements were performed, i.e., 2 samples for each concentration of each solvent were collected, and each sample was exposed twice to the electronic nose. The experiments were repeated for many weeks over repeated batches for a period of several months.

The testing was performed using a 5-fold cross validation technique for each week. We retained the data collected in a day as test set and used the remaining data as training set. The test day was rotated so that each weekday was used as test set. Tables I, II and III summarise the most representative results obtained respectively with Fuzzy Logic, Evidential Logic and Neural Module.

		Week 1	week 2	week 3
Train	average	83.1	82.5	82.8
	std. dev.	3.8	0.8	5.8
Test	average	80.0	81.7	82.1
	std. dev.	8.1	4.8	2.4

Table I *Fuzzy Logic Model:* Percentage of successful classifications on 5 day-fold cross-validation for three consecutive weeks.

		Week 1	week 2	week 3
Train	average	92.8	87.2	94.5
	std. dev.	0.6	0.5	0.5
Test	average	79.6	76.3	70.4
	std. Dev.	6.4	5.7	25.6

Table II *Evidential Logic Model:* Percentage of successful classifications on 5 day-fold cross-validation for three consecutive weeks, using the MA-ID3 method

		Week 1	week 2	week 3
Train	Average	94.1	88.5	90.4
	std. Dev.	3.4	4.3	3.1
Test	Average	85.0	82.9	85.8
	std. Dev.	6.5	2.7	10.1

Table III *Neural network Classification Model:* Percentage of successful classifications on 5 day-fold cross validation for three consecutive weeks.

We observed that, when using consisting datasets where the three classifiers had a "stable" behaviour (i.e. success results for the training stage were not much different from the ones obtained in the test stage), the "Merging" strategy between the modules themselves always produced classification rates higher than the best of the three methods. Also, in the end of the measurement activity with packaging material, and in order to assess the ability of the INTESA prototype to recognise odorants along the time, we performed the following experiment: we used the first weeks as training set and the remaining weeks as test set, working with datasets including measurements taken along periods of several weeks. In this case, the average recognition accuracy obtained by the INTESA integrated software was in the same range (85% ± 5%) as the accuracy obtained week by week, while the classification rate obtained by a conventional DFA classifier dropped at a level close to 65%, in contrast with a result similar to the INTESA software that was obtained by DFA with cross-validation, week-by-week.

5. Conclusions

In this paper, the application of a novel processing strategy for Chemical Sensor Array Data has been described. It has been demonstrated that "Intelligent" techniques used by the "INTESA" processing software show interesting results vs. conventional techniques such as DFA, when applied to large datasets acquired over significant time spans. Moreover, the integration of different processing techniques used by the INTESA package have demonstrated itself to be successful in most cases during the experimental activity.

The above results are particularly encouraging considering a real industrial application of the INTESA processing approach.

References

Baldwin, J. F., Martin, T. P. and Pilsworth, B. W. (1988). "FRIL Manual (Version 4.0)". Fril Systems Ltd, Bristol Business Centre, Maggs House, Queens Road, Bristol, BS8 1QX, UK. 1-697.

Baldwin, J. F. (1992). "The Management of Fuzzy and Probabilistic Uncertainties for Knowledge Based Systems" in Encyclopedia of AI, Ed. S. A. Shapiro, John Wiley. (2nd ed.) 528-537.

Baldwin, J. F., Martin, T. P. and Pilsworth, B. W. (1995). "FRIL - Fuzzy and Evidential Reasoning in AI", Research Studies Press (John Wiley).

Baldwin, J. F., Lawry, J. and Martin, T. P. (1997). "A mass assignment based ID3 algorithm for decision tree induction." Intl. Journal of Intelligent Systems 12(7): 523-552.

Baldwin, J. F., Martin, T. P. and McCoy, S. A. (1998). "Incremental Learning in a Fril-based Odour Classifier", Proc. EUFIT-98, Aachen, Germany, 1216-1220.

Gardner J.W., 1991,"Detection of Vapours and Odours from a Multisensor Array Using Pattern Recognition - Part 1. Principal Component and Cluster Analysis", Sensors and Actuators B, Vol.4, pp. 109-115.

Gardner J.W., E.L. Hines, H.C. Tang, 1992,"Detection of Vapours and Odours from a Multisensor Array Using Pattern Recognition techniques - Part 2. Artificial Neural Network", Sensors and Actuators B, Vol.9, pp. 9-15.

H. T. Nagle, 1997, "Gas Sensor Arrays and the Electronic Nose", IEEE- electronic nose workshop (Enose'97), Marriott Marquis Hotel, Atlanta, GA, Nov. 7.

Lazzerini, B., Maggiore, A., Marcelloni, F., "Classification of odour samples from a multisensor array using new linguistic fuzzy method", Electronics Letters, Vol. 34, No. 23 (November 1998), pp. 2229- 2231

L.Khodja, Foully L., Benoit E., Talou T., "Fuzzy techniques for coffee flavour classification", Proceedings of IPMU 96, 1996, pp. 709-714.

Nagle H.T., Gutierrez-Osuna R., Schiffman S.S., 1998, "The How and Why of Electronic Noses", IEEE Spectrum, September.

S. S. Schiffman, B. G. Kermani, and H. T. Nagle, 1996, "Use of an Electronic Nose to Detect the Source of Environmental Pollutants", Pap. 497, Pittcon '96, Chicago IL, Mar 3-9.

Savitzky, A., Golay, J, 1964, "Smoothing and Differentiation of Data by Simplified Least Squares Procedures". Analytical Chemistry no 36, p. 1627.

Singh S., Hines E.L., Gardner J.W., 1996, "Fuzzy neural computing of coffee and tainted-water data from and electronic nose", Sensors and Actuators B, Vol.30, pp. 185-190.

Strackeljan J., D. Behr, T. Kocher, 1997, "Fuzzy-pattern recognition for automatic detection of different teeth substances", Fuzzy Sets and Systems, No. 85, p. 275.

Section 3: Data Processing
Paper presented at the Seventh International Symposium on Olfaction and Electronic Noses, July 2000

107

An Electronic Nose for Recognizing Combustible Gases using Thick Film Sensor Array and Neural Network

Dae-Sik Lee[1], Jeung-Soo Huh[2], Hyun-Gi Byun[3], and Duk-Dong Lee[1]

[1]*School of Electronics and Electrical Eng. Kyungpook National University, Taegu, Korea*

[2]*Dept. of Metallurgical Eng. Kyungpook National University, Taegu, Korea*

[3]*Dept. of Information and Communication Eng. Samchok National University, Samchok, Korea*

E-mail ddlee@ee.knu.ac.kr

Abstract : We propose an electronic nose for recognizing combustible gases using a SnO_2 based thick film sensor array and neural network for pattern recognizer, within the ranges of TLVs and LELs. The electronic nose can not only identify the kinds of the combustible gases but also recognize concentration values of the identified gas. A sensor array with ten discrete sensors was developed with the aim of recognizing the kinds and quantities of combustible gases, such as methane, propane, butane, and carbon monoxide within the standard ranges. The sensor array consisted of gas-sensing materials of SnO_2, plus a heating element based on a meandered platinum layer. The sensors on a sensor array were designed to produce a uniform thermal distribution and show a high and broad sensitivity to low concentrations of about 100 ppm. Using the sensing signals of the array along with an artificial neural network, an electronic nose system was then implemented. The characteristics of the multi-dimensional sensor signals obtained from sensors were analyzed using the principal component analysis (PCA), and a gas pattern recognizer was implemented using a neural network with an error-back-propagation learning algorithm. The simulation and experimental results demonstrated that the proposed electronic nose was effective in identifying combustible gases. For real time processing, a digital signal processor (DSP) board was then used to implement the electronic nose.

1 Introduction

Recently, as the usage of LNG and LPG gas has increased, the frequency of accidental explosions due to leakage has sharply increased. The ability to detect and precisely measure gas leakage is crucial in preventing the occurrence of such accidents. Accordingly, the development of sensors and systems that can selectively detect and determine various kinds and the quantities of the combustible gases within the ranges of the explosion limits (LELs) and threshold limit values (TLVs) is urgently needed. Until now, there have been various reports on the use of different kinds of metal oxide sensors for detecting combustible gases [1]. Furthermore, to overcome the poor selectivity and long-term drift of a sensor signal, studies have been worked on development of an electronic nose with sensor array systems consisting of various sensing materials such as metal oxides, electrochemical sensors, conducting polymers, surface acoustic wave devices, and hybrid sensor arrays [2-6].

For continuous use, without relearning, an electronic nose requires high reliability, stability, and sensitivity from the individual sensors that make up the nose. However, sensor responses tend to drift significantly when sensors are used over a long time. It is needed to fabricate more reliable and stable sensors, or adjust the pattern recognition routine according to variations in the sensor response. Accordingly, nano-sized tin oxide materials were synthesized using the coprecipitation method and ten sensors were fabricated on a substrate.

The characteristics of multi-dimensional sensor signals obtained from 10 sensors were analyzed using the PCA, and we developed an electronic nose system that can not only classify the kinds of the combustible gases but also provide the concentration values of the identified gas. Above all, we adapted a multi-layer neural network with an error-back-propagation learning algorithm for identification of the kinds of gas and for quantifying of gas concentration. Using the sensing signals of the arrays and neural networks, a gas pattern recognition was then implemented using a DSP board with the aim of real-time classifying and quantifying the combustible gases, including butane, propane, methane, and carbon monoxide below their explosion limit values (TLVs).

2 Experimental

It has been previously reported that nano size effect through Ca catalyst along with the spillover effect caused by Pt catalyst enhance the sensitivity of sensing materials to gases even at low concentrations [8]. For preparing the raw materials, a solution containing $SnCl_4$,

$(CH_3CO_2)_2Ca \cdot xH_2O$ and, $H_2PtCl_6 \cdot 6H_2O$ was coprecipitated by controlling the pH. The

powder was then washed, dried, and calcined in an electric furnace at 600□ for 1h. The Ca and Pt contents were both set at 0.1 wt.% relative to the amount of $SnCl_4$. Using this powder as a base material, ten kinds of sensing materials with different additives such as Pd, Au, Pt, La_2O_3, CuO, Sc_2O_3, TiO_2, WO_3, ZnO, and V_2O_5 within a range of 0.1~5 wt.% were evenly mixed to modify the sensitivity spectrum of the sensors [9].

The sensing films were all prepared using the silk printing technique. After being simultaneously deposited, the sensing films were then annealed at 800□ for 2 h in air. The following presents a brief outline of the results produced by the sensors on the array. All sensing materials were deposited on a substrate (1.1×1.2 cm^2) with a heating element and a temperature sensor. Platinum pads and contacts were deposited on the front-face of the substrate. The structure of a sensor array is shown in Fig. 1(a). The sensor array was connected with the long bonding pins for electrical conduction and suspended in air for thermal isolation from socket and peripheral circuit. The sensor array mounted on a printed circuit board is shown in Fig. 1 (b).

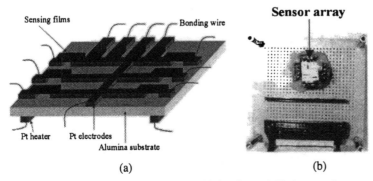

Figure 1. Structure of sensor array; (a) drawing and (b) photograph.

The characteristics of the sensor array were then tested in a testing chamber after injection of combustible gases. The sensor signals were transferred to a test system and monitored using a personal computer in a testing chamber.

A voltage detecting method was used to calculate the sensitivity of the sensor [10]. The sensitivity was defined as $(R_{air}-R_{gas})/R_{air} \times 100$ (%) where R_{gas} and R_{air} are the electrical resistances in combustible gas and clean air, respectively.

3 Results and discussions

3.1. Characteristics of SnO$_2$-based thick film sensors for combustible gases

To selectively determine the quantities of combustible gases within low explosion limits (methane: 5.3 vol.%, propane: 3.3 vol.%, butane: 1.9 vol.%, and carbon monoxide: 12.5%) and threshold limit values (butane: 800 ppm and carbon monoxide: 50 ppm), stable sensors with a high sensitivity to low concentrations are required. To improve the sensitivity of a sensor, the particle size, specific surface area, and crystalline structure are all recognized as significant parameters [11]. Accordingly, the microstructure of various base materials was investigated for its adaptability. As a result, The SnO$_2$/Ca,Pt base material exhibited a nano-scaled particle size, a high specific surface area, and a tetragonal structure after being calcined at 600□ for 1h [8].

The response time of a sensor to a gas is directly related to gas detection and classification.

The time response curve of the sensor array to 1000 ppm propane at 400☐ is shown in Fig. 2. It shows that the gas reaction began to saturate within 3~5 seconds and desorption of the gas was also completed within 30 seconds. The detection and recognition of a gas was thus completely finished within 2~3 minutes, in real time process. The sensors also exhibited various gas-sensing curves according to the kinds of additives.

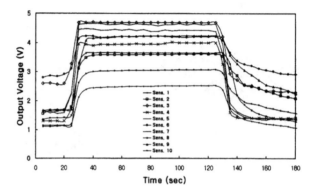

Fig. 2. Time response of sensors with 1000 ppm propane gas.

The gas-sensing ability of the component sensors was evaluated to determine the operating ranges of a sensor array. The sensitivity of the SnO₂-based thick film sensors to combustible gases (methane, propane, butane, and carbon monoxide) in a range of 0☐5,000 ppm concentration at 400☐ is investigated. The sensors showed good sensitivity and good linearity to combustible gas concentration. All SnO₂ thick film sensors showed good sensitivity to butane, propane, carbon monoxide and methane in order. It shows that the sensitivity of sensors in array start to be saturated at about 1,000 ppm gas concentration and is fully saturated near 5,000 ppm level, and hence it is difficult to obtain a variety of sensing patterns in the concentration range of over 3,000 ppm. Accordingly, the available ranges of this sensor array to recognize the kinds and quantities of combustible gases can be determined as 0☐3,000 ppm level in this electronic nose system.

Differentiated sensing patterns for the sensors are critical for conferring selectivity to a sensor array. All sensors shows a variety of sensitivity ranging from 20 to 100 % to the tested combustible gases within the ranges of their LELs and TLVs. This figure also shows the effects of additives in the base material, that is, sensing materials with the addition of CuO showed the highest sensitivity to all combustible gases measured in this experiment. Accordingly, it was believed that these additives like noble metals or oxides could confer SnO₂ base material to the differentiated sensitivity patterns.

The possibility of classifying the kinds and quantities of simple/mixed combustible gas can be achieved by using a PCA. After the data obtained from the experimental results were measured by four repetitive experiments, they were transposed onto two axes, as shown in Fig. 3. This demonstrates the ability to cluster points roughly in types according to their species and quantities. However, it was difficult to clearly classify the 1000 ppm propane and 500 ppm butane as they are located close each other. Therefore, to precisely classify clusters with similar characteristics, a gas pattern recognizer using a multi-layer neural network was employed along with an error-back-propagation learning algorithm, which is known to be effective in classifying and identifying complex nonlinear patterns [12].

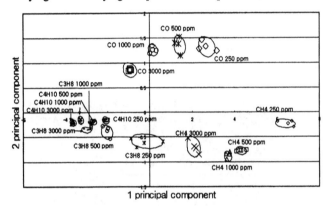

Figure. 3. PCA results of sensing data obtained from sensor array.

3.2. Implementation of gas pattern recognizer and gas recognition system

The neural network was implemented to identify the kinds of target gas and to recognize the continuous concentration value of it. Using the sensitivity signals from the sensor array as multi-dimensional input patterns, a gas pattern recognizer using a multi-layer neural network with an error back propagation learning algorithm was then implemented []. The neural network consisted of an input layer with ten nodes that received the data from the sensors on the sensor array, a hidden layer with eighth neurons, and an output layer with sixteen nodes. The 16 nodes in the output layer indicated the recognition results for four concentration levels of four kinds of combustible gases; butane, propane, methane and carbon monoxide. The learning process was repeated until the learning error reached 0.0001.

In this study, the experimental data was measured with 250, 500, 1000, and 3000 ppm concentrations of four combustible gases; butane, propane, methane, and carbon monoxide. The developed gas classifier could precisely identify the correct species of all experimental gases, which include the gases with the concentration levels used in learning and those not used in learning. In order to construct the gas concentration recognizer, the following equation is designed by using output of neural network.

$$\text{Max}\left[\frac{\sum_{i=SG} out(i) \times corres\_concentration(i)}{\sum_{i=SG} out(i)}\right] \text{ for all } S_G \qquad (1)$$

Where S_G is an output node in neural network corresponding same kinds of gases, i.e. $S_{methane}$, $S_{propane}$, S_{butane}, and S_{CO}, $out(i)$ is i-th output value obtained from the neural network, and is concentration corresponded to i-th output node. Fig. 4-17 shows, as an example, the some results in case of injecting propane for 4 times repeatedly at certain concentration levels of 250, 375, 500, 750, 1,000, 2,000 and 3,000 ppm, respectively. It indicated relatively precise concentration value within 3,000 ppm. It also shows similar results to the other combustible gases. Therefore, the combustible gas recognizer using neural network can discriminate the five combustible gases selectively and can indicate the quantities of simple/mixed gases approximately in a range of 0□3,000 ppm.

Figure. 4. Measured concentration versus real concentration for propane gas

A fabricated combustible gas recognition system consists of a sensor array, a data acquisition board, a power supplier, and DSP board. An analog multiplexer, a low-pass filter, and a signal transformer for level shifting were all included in a data acquisition board for transferring the signal to a DSP board. The learned connection-weights were downloaded in the DSP board memory. For real-time monitoring the combustible gases, the recognized gas species and concentrations can be displayed on a liquid crystal display panel.

4   Conclusion

An electronic nose system for monitoring the combustible gases, with a sensor array and a neural network pattern recognizer, was implemented in hardware using a DSP board.

To recognize the various kinds and quantities of combustible gases within the ranges of their TLVs and LELs, a sensor array with ten $SnO_2/Ca,Pt$–based metal oxide sensors integrated on an alumina substrate with size of 1.1×1.2 cm^2 was fabricated. To give selectivity to the sensors, several additives, like Pt, Pd, Au, CuO, La_2O_3, Sc_2O_3, TiO_2, WO_3, or ZnO, were mixed with the base material. The resulting sensor arrays showed high and selective sensitivities to low gas concentrations at 400□. A neural network analyzer with an error-back-propagation learning algorithm was implemented to classify the kinds of gas and to recognize the continuous values of gas concentration.

A DSP board was utilized in a portable system for monitoring combustible gases, such as butane, propane, methane, and carbon monoxide. The fabricated recognition system could not only classify the gas species precisely but also identify the continuous concentration value of each combustible gas in real-time process.

Acknowledgements: This work was supported by Korea Government funds and electronic technology center at Kyungpook National University, Korea.

References

[1] D.-D. Lee, Chemical Sensor Technology, Vol. 5, Kodansha Ltd. 1994, 79-99

[2] J.W. Gardner and P.N Bartlett, Ed., Sensors and sensory systems for an electronic nose, NATO advanced research workshop, Reykjavik 1991

[3] U. Weimar, K.D. Schierbaum, W. Gopel, Sensors and Actuators, B1, 1990, 93-96

[4] S. Somov, G. Reinhardt, W. Gopel, Sensors and Actuators, B 35-36, 1996, 409-418

[5] K.C. Persaud, S.M. Khaffaf, J.S. Payne, D.H. Lee, H.G. Byun, Sensors and Actuators, B35-36 (1996) 267-273

[6] C.-F. Mandenius, T. Eklov, I. Lundstrom, Biotechnology and Bioengineering 55(2) 1997 427-238

[7] M. Lee, S.- Y. Lee, and C.-H. Park, Journal of Intelligent and fuzzy systems 2 1994, 1-14

[8] Y.H. Hong, D.D. Lee, J. of the Korean Sensors Soc. 4(2) 1995 31-44

[9] D.-S. Lee, H.-Y. Jung, J.-W. Lim, S.-W. Ban, M. Lee, J.-S. Huh, D.-D. Lee, Sensors and Actuators B, 2000 (in press)

[10] T. Seiyama, Denki Kagaku, 40 (1972) 244-250.

[11] N. Yamazoe, N. Miura, Chemical Sensor Technology, Vol.4, Kodansha Ltd., Tokyo, Japan, (1992) 19-42.

[12] S. Haykin, Neural networks; a comprehensive foundation, Macmillan college publishing Co., New York, USA, 1994

Section 3: Data Processing
Paper presented at the Seventh International Symposium on Olfaction and Electronic Noses, July 2000

115

Application of a Multilayer Perceptron Based on the Levenberg-Marquardt Algorithm to Odour Pattern Classification and Concentration Estimation using Odour Sensing System

Jeong-Do Kim[1], Hyung-Gi Byun[2], Krishna C. Persaud[3]

[3]Dept. of Control and Instrumentation Eng. Samchok National University, Samchok, Korea

[2]Dept. of Information and Communication Eng. Samchok National University, Samchok, Korea

[3]Dept. of Instrumentation and Analytical Science, UMIST, PO Box 88, Manchester, UK

E-mail byun@samchok.ac.kr

Abstract : This paper presents the application of multilayer perceptron baed on the Levenberg-Marquardt algorithm to odour pattern classification and concentration estimation using odour sensing system. We adapted the Levenberg-Marquardt (L-M) algorithm having advantages both the steepest descent method and Gauss-Newton method instead of the conventional steepest descent method for the simultaneous classification and concentration estimation of odours. It is confirmed that the multilayer perceptron based on the L-M algorithm is capable of not only the identification of odour classes, but also the prediction of odour concentrations for volatile chemicals with different concentration levels simultaneously throughout the experimental trails.

1. Introduction

A variety of pattern recognition techniques may be utilized for odour pattern classification. These include artificial neural networks that take the input patterns generated by the array of conducting polymer sensors [1], which may be trained to associate these patterns with particular classes of volatile chemicals that may be of interest to the user. Such architectures include 2-layer systems trained by conventional back-propagation of error [2] and radial basis function network [3]. Of particular interest to us is not only the identification of odour classes, but also the prediction of odour concentrations, even when the background may be complex. The quantification of odours is very desirable feature in real life and is much more difficult to predict concentration levels of single chemicals or mixture than classification of different chemicals. Most of researchers have tried to classify the odour identification and to predict concentration levels for odours separately using different algorithms.   In this paper, we have investigated the properties of multilayer perceptron (MLP) for odour pattern classification and concentration estimation simultaneously.   When the MLP may be has a fast convergence speed with small error and excellent mapping ability for classification, it can be possible to use for classification and concentration prediction of volatile chemicals simultaneously.   However, the conventional MLP, which is back-propagation of error based on the gradient method, was difficult to use for odour classification and concentration estimation simultaneously, because it slow to converge and may fall into the local minimum.   We adapted the Levenberg-Marquardt (L-M) algorithm [4,5], having advantages both the steepest descent method and the Gauss-Newton method instead of the conventional steepest descent method for the simultaneous classification and concentration estimation of odours.   It is confirmed by experimental trails with the electronic odour sensing system, which has been constructed at UMIST by Dr. Krishna C. Persaud.

2. The Approach Method for Adaptation Algorithm

There are a number of algorithms for error optimization.   Figure 1 illustrates the topology of optimization algorithms.

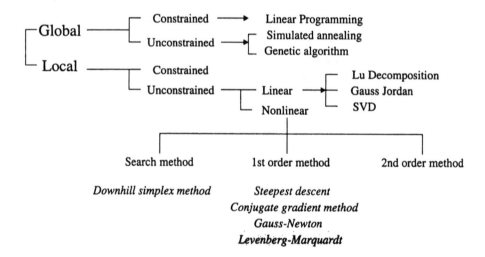

Figure 1. The topology of optimization algorithms

The best known non-linear learning optimization algorithms for artificial neural network can be shown figure 1 such as steepest descent method, Gauss-Newton method, and Levenberg-Marquardt method.   The optimization characterizations of those algorithms can be compared with Table 1, which were confirmed by many researchers [4,5].

	EFFICIENCY	CONVERGENCE	COMPLEXITY	DERIVATIVES
Steepest descent	O	X	O	1st order
Gauss-Netwon	OO	X	O	1st order
Levenberg – Marquardt	OO	O	O	1st order
Newton	OO	X	X	2nd order

Table 1. The comparison for learning optimization methods for artificial neural network

We apply a multilayer perceptron based on the Levenberg-Marquardt algorithm to odour pattern classification and concentration estimation having basic configuration shown in figure 2.

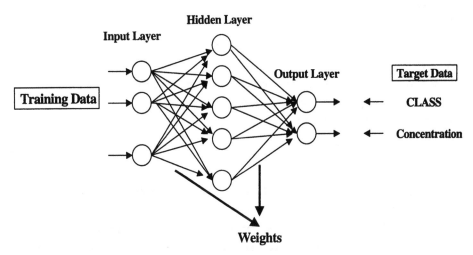

Figure 2. The Multilayer perceptron network basic configuration for classification and concentration estimation.

The updating rule for MLP based on the L-M algorithm is as follows [4,5].

$$\omega_{m+1} = \omega_m - [J^T(\omega_m)J(\omega_m) + \lambda_m I]^{-1} J^T(\omega_m)e(\omega_m)$$

Where J is the Jacobian matrix of derivatives of each weight, λ is a scalar, and e is an error vector. $J^T J$ is called the Hessian matrix and $J^T e$ is the gradient. The learning is carried out by steepest descent method when λ is large, the Gauss-Newton method is used for learning when λ is small.

Using the above updating rule, the MLP based on L-M algorithm is carried out learning procedure as follows:

Object function :

$$F(\omega) = \sum_{i=0}^{N} e_i^2(\omega)$$

Where N is the number of samples.

Algorithm
1. $\lambda \leftarrow \lambda_0$: Initial value for λ; Weights are initialized.
2. Compute $F(\omega_m)$
3. Compute Jacobian J_k
4.

$$\omega_{m+1} = \omega_m - [J^T(\omega_m)J(\omega_m) + \lambda_m I]^{-1} J^T(\omega_m) F(\omega_m)$$

5. Compute $F(\omega_{m+1})$
6. If $F(\omega_{m+1}) > F(\omega_m)$ then $\lambda \leftarrow \theta \times \lambda$ and go to 4. Where θ is a constant with $1 < \theta < 10$.
7. If $F(\omega_{m+1}) < SSE(Square Sum Error)$ goal then stop.
8. $\lambda \leftarrow \alpha \times \lambda$, where α is divider.
9. $\omega_m \leftarrow \omega_{m+1}$ go to 2.

3. Results

The capability of MLP based on the L-M algorithm, which is developed for classification and concentration estimation for volatile chemicals simultaneously, is tested against real experimental data collected with odour sensing system using conducting polymer sensor array. The response patterns obtained from the experimental measurement were prepared for training and testing sets. The training sets were prepared from the five different volatile chemicals according to concentration levels, and the testing sets were also prepared by 5 patterns throughout the measurement. The training sets, which were used for experimental measurement, were shown Table 2.

The figure 3 showed the classification and concentration estimation result for training data sets using the MLP based on L-M algorithm. The figure 4 illustrated test data sets, which were not seen in the training session, were well classified by class and predicted concentration levels simultaneously using the previously trained weights.

ETHANOL	METHANOL	BUTANONE	N-BUTYL ACETAE	MIXTURE
2200ppm (3 patterns)	3200ppm (3 patterns)	3700ppm (3 patterns)	1000ppm (3 patterns)	2700ppm (3 patterns)
4500ppm (3 patterns)	6500ppm (3 patterns)	7300ppm (3 patterns)	2000ppm (3 patterns)	3900ppm (3 patterns)
6700ppm (3 patterns)	9700ppm (3 patterns)	15000ppm (3 patterns)	4000ppm (3 patterns)	7200ppm (3 patterns)
11200ppm (3 patterns)	12900ppm (3 patterns)	18000ppm (3 patterns)	5000ppm (3 patterns)	13000ppm (3 patterns)

Table 2. Training data sets according to concentration levels. The data set 'mixture' consisted of a range of ethanol concentrations in a constant methanol background.

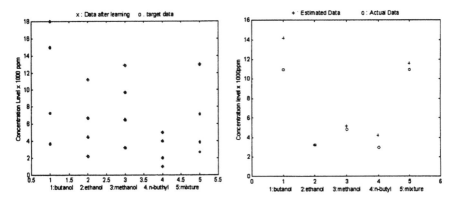

Figure 3. Training Results   Figure 4. Testing Results

4. Discussion

This paper presents the application of multilayer perceptron baed on the Levenberg-Marquardt algorithm to odour pattern classification and concentration estimation using odour sensing system. Throughout the experimental trails, it is confirmed that the MLP based on the L-M algorithm is capable of not only the identification of odour classes, but also the prediction of odour concentrations for volatile chemicals with different concentration levels simultaneously. These preliminary results are promising that MLP based on the L-M algorithm is applicable to odour sensing systems for environmental monitoring, cosmetic, and beverage industries.

Acknowledgements: This work was supported financially by Zenith Information & Communication Co. Ltd., Samchok National University Internal Research Fund, Korea.

References

[1] Persaud KC and Travers P 1991 Intelligent Instruments and Computer Vol. 147.

[2] Rumelhart D, Hinton GE, Williams RJ 1986 Nature Vol.323:533-536

[3] Lee D-H, Payne JS, Byun H-G, Persaud KC 1996 Lecture Notes in computer science Vol. 1112:299-304

[4] Master T 1995 Advanced algorithms for neural networks, Wiley 47-71

[5] Finschi L 1996 Clausiusstrasses 45 CH-8092 Zuerich

Section 3: Data Processing
Paper presented at the Seventh International Symposium on Olfaction and Electronic Noses, July 2000

121

Application of Adaptive RBF Networks to Odour Classification using Conducting Polymer Sensor Array

Hyung-Gi Byun[1], Nam-Yong Kim[1], Krishna C. Persaud[2], Jeung-Soo Huh[3], Duk-Dong Lee[4]

[1]*Dept. of Information and Communication Eng. Samchok National University, Samchok, Korea*

[2]*Dept. of Instrumentation and Analytical Science, UMIST, PO Box 88, Manchester, UK*

[3]*Dept. of Metallurgical Eng. Kyungpook National University, Taegu, Korea*

[4]*School of Electronics and Electrical Eng. Kyungpook National University, Taegu, Korea*

E-mail byun@samchok.ac.kr

Abstract : This paper presents the application of adaptive RBF(Radial Basis Function) Networks to odour classification using conducting polymer sensor array. The performance of RBF Network is highly dependent on the choice of centers and widths in basis function. For the fine tuning of centers and widths, those are initialized by ill-conditioned genetic fuzzy c-means algorithm and the distribution of input patterns in the very first leaning stage, the stochastic gradient (SG) method is adapted. The adaptive RBF Network, which has tuned centers and width by SG method, has shown good classification performance for complex and mixture chemical patterns and confirmed by experimental trails.

1. Introduction

The electronic odour sensing system, which had been constructed at UMIST by Dr.Persaud, comprised an array of conducting polymer sensors mounted on a ceramic substrate together with associated electronics. Extremely selective information for discrimination between adsorbed chemical species or mixtures can be obtained by analysis of the cross-sensitivities between sensor elements. The relative responses between sensor elements produce patterns that may be unique 'fingerprints' that may be used as odour descriptors. This strategy has been successful in the design of chemical sensors that are capable of detecting some volatile chemicals that are difficult to detect by other methods. The modulation of electrical conducting polymers by external physical and chemical interactions make them attractive use in chemical sensing [1]. When odour is presented, all the sensors respond with a reverse change of resistance. The intensity of the response is dependent on the affinity of chemical species for individual sensors, and is proportional to concentration present. If the steady-state response of each sensor to the absolute sum of the response of the entire array is taken, then the raw data can be transformed into a pattern, that is unique to that particular chemical species, or mixture, and can be used as a 'fingerprints' to identify it. Figure 1 showed the relative response patterns obtained from the normalization of raw data from the sensors.

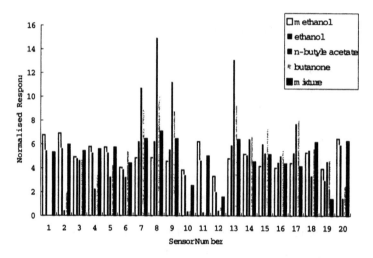

Figure 1. The relative response patterns from the normalization of raw data from the sensors. Patterns obtained for methanol, ethanol, n-butyl acetate, butanone ,and a mixture of ethanol and methanol are shown.

Using electronic odour sensing system, it is desirable to discriminate between chemicals, and compare one sample with another. The ability to classify pattern characteristics from relatively small pieces of information has led to growing interest in methods of sensory recognition.

We have investigated the characteristics of radial basis function (RBF) networks applied to odour classification problems. The RBF Network is to design an artificial neural network with good generalization ability and to train rapidly, usually orders of magnitude faster than error back-propagation, while exhibiting none of back-propagation's training pathologies such as paralysis or local minimum problems. The architecture of RBF Network has to be simple and consist of input, hidden, and output layers. Each node in hidden layer is calculated by a non-linear function on the incoming input and output, and the output is simply weighted linear summation of the non-linear function. The output, which is called a one-dimensional feature and extracted by the RBF Network, has to clearly visualize the non-linear projection. The RBF Network may be expected to superior to the other learning algorithm, particularly in terms of processing speed and solvability of non-linear pattern responses in odour analysis. Basically, the mapping function implemented by an RBF Network differs from the mapping function used with back-propagation. The principal difference is in the hidden layer units. A RBF network hidden layer unit has one parameter for each input. These parameters are not weights placed on the inputs; rather, they are the coordinates in input space of a point that is the center of the hidden unit's output function. The basis functions in hidden layer produce a localized response to input and typically uses hidden layer units with Gaussian response function [2]. Therefore, the performance of RBF network as a classifier is highly dependent on the choice of centers and widths in basis function. Throughout our previous publications [3,4,5], it has been confirmed.

Most of the learning algorithms for RBF networks have divided the two stages processing. Firstly, a clustering method such as the k-means algorithm or fuzzy c-means algorithm or genetic fuzzy c-means algorithm is applied to the input patterns in order to determine the centers for hidden layer units. After the centers are fixed, the widths are determined in a way that reflects the distribution of the centers and input patterns. Once the centers and widths are fixed, the weights between hidden and output layer are trained by a single shot process using singular value decomposition (SVD) or by the least-mean-squares (LMS) algorithm. This two-stage method provides some useful solutions in pattern classification problem. However, since the centers and widths are fixed after they are chosen and only the weights are adapted for supervised learning, this method often results in not satisfying performance when input patterns are not particularly clustered. In the our previous publications, we found that the genetic fuzzy c-means algorithm was provided relatively good performance compared with another center selection method. However, it is still difficult to optimum centers from various odour patterns, especially the genetic fuzzy c-means algorithm was difficult to determine the various parameters which were used and the width was not considered except experimental trails.

In this paper, we proposed an adaptation method of centers and widths for RBF Networks. It shows the better performance compared with the previous classification results without centers and widths adaptation.

2. The Approach Method

For a given set of input patterns measured using a twenty or thirty-two element sensor array, the genetic fuzzy c-means algorithm with random initial parameters is carried out to find locations of clusters' centers which are then fed into the hidden layer units of RBF Network. The Euclidean distance between the input patterns and the clusters' centers is evaluated, then a basis function such as Gaussian function with initial widths is applied. The weights between hidden and output units are trained by a single shot process using the SVD method, because RBF Network is applied as a supervised learning algorithm for odour classification. Though the SVD provides excellent function for weights calculation, the centers and widths are required to be finely tuned for better performance. For the tuning of centers and widths, those are initialized by ill-conditioned genetic fuzzy c-means algorithm and weights are also initialized by SVD in the very first learning stage, the stochastic gradient (SG) method is adapted to finely tune centers and widths as follows [6]:

$$\Delta\sigma_j^{(n)} = \sigma_j^{(n+1)} - \sigma_j^n = -\mu_s \frac{\partial e_n^2}{\partial \sigma_j^{(n)}} = \mu_s e_n \omega_j^{(n)} \exp(\frac{-\left\|x_n - c_j^{(n)}\right\|^2}{(\sigma_j^n)^2}) \frac{\left\|x_n - c_j^{(n)}\right\|^2}{(\sigma_j^n)^3}$$

$$\Delta c_j^{(n)} = c_j^{(n+1)} - c_j^n = -\mu_c \frac{\partial e_n^2}{\partial c_j^{(n)}} = \mu_c e_n \omega_j^{(n)} \exp(\frac{-\left\|x_n - c_j^{(n)}\right\|^2}{(\sigma_j^n)^2}) \frac{(x_n - c_j^{(n)})}{(\sigma_j^n)^2}$$

where μ_s and μ_c are adaptation coefficients for widths σ_j, and centers c_j, respectively, and they control the speed of adaptation. Note that in widths tuning formula the adaptation is with respect to σ_j rather than to σ_j^2, in order to prevent σ_j^2 from becoming negative during learning.

3. Results

The adaptive RBF network, which has tuned centers and widths by SG method, has showed good classification performance for the complex and mixture chemical patterns and confirmed the SG method has made fine tuning given relatively ill-conditioned clustering centers and widths. It was confirmed by the experimental trials.

The figure 1 showed the 1-D visualization classification result of five volatile chemicals obtained from presenting methanol, ethanol, n-butyl acetate, butanone, and mixture of ethanol and methanol using RBF Networks based on ill-conditioned genetic fuzzy c-means algorithm center selection technique with one center from each class, when tested against very noisy data from conducting polymer sensor array. The data set 'mixture' consisted of a range of ethanol concentrations in a constant methanol background. The data sets 'ethanol' and 'mixture' produced poor classification results as shown in the figure 1. The figure 2 illustrated the classification result using adaptive RBF network based on the SG tuning method for centers and widths with one center from each class. It showed the improvement of classification prediction results compared with the previous result in the figure 1. The data sets 'ethanol' and 'mixture' are classified with much better accuracy using adaptive RBF network

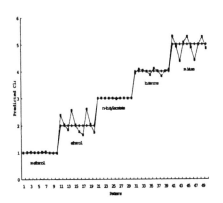 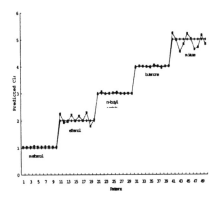

Figure 2. Classification Result without SG.   Figure 3. Classification Result with SG.

4. Discussion

This paper presents adaptive RBF Networks to complex odour classification using SG method for tuning of centers and widths. Throughout the experimental trails, we confirmed that adaptive RBF Network based on genetic fuzzy c-means algorithm and SG tuning technique showed good classification performance compare with RBF Network based on genetic fuzzy c-means algorithm only, and have found wide applicability for complex odour classification problems. It also may be a useful tool for 1-D visualization of pattern classification from array-based chemical sensing system.

Acknowledgements: This work was partly supported by various Korea Government funds and electronic technology center at Kyungpook National University, Korea.

References
[1] Pelosi P and Persaud KC 1988 NATO ASI Series Vol. F43: 49-70
[2] Moody J and Darken C 1989 Neural Networks Vol.1 no.2 91-107
[3] Lee D-H, Payne JS, Byun H-G, Persaud KC 1996 Lecture Notes in computer science Vol.

1112:299-304

[4] Persaud KC and Byun H-G 1998 Industrial applications of neural networks (World Scientific) 85-90

[5] Byun H-G, Persaud KC, Kim J-D, Lee D-D 1999 Proceedings of ISOEN'99: 237-240

[6] Cha I and Kassam SA 1995 IEEE Journal on Selected Areas in Communications, Vol. 13 No. 1 122-131

Section 3: Data Processing
Paper presented at the Seventh International Symposium on Olfaction and Electronic Noses, July 2000

127

Fuzzy logic based classification of olive oils

Glauco Bargagna†, Beatrice Lazzerini‡[1] and Ashton Partridge§

† I.S.E. Ingegneria dei Sistemi Elettronici s.r.l., Vecchiano (Pisa), Italy

‡ Dipartimento di Ingegneria della Informazione, University of Pisa, Via Diotisalvi 2, Pisa, Italy

§ Industrial Research Limited, Gracefield Road, Lower Hutt, New Zealand

Abstract. In this paper we present an electronic nose for olive oil classification. The electronic nose employs 8 conducting polymer (CP) sensors. The sensor signals are processed by a pattern recognition system based on fuzzy logic.

1. Introduction

In the last years, a growing interest in mimicking the mammalian olfactory system has arisen with the aim of developing a computer-based system which is able to classify odours and perhaps to measure odour intensity and nature. Such a system is called an *electronic nose* [1]. Typically, electronic noses integrate an array of a few sensors with partially overlapping sensitivities to odours, and a pattern recognition system [2]. Various types of sensor technologies as well as pattern recognition techniques are available. Non-linear pattern-recognition methods have proven more successful than linear ones thanks to their ability to handle the highly non-linear transduction properties of the sensors and to compensate for sensor drift.

In this work, we present a fuzzy logic based method for classification of olfactory signals. We explicitly model the uncertainty present in sensor responses building a fuzzy model of the 'typical' response of a given sensor to an odour class [3]. The model, called *shape model*, describes the shape of the sensor response to the odour and is obtained considering a set of responses of the sensor to the odour in repeated experiments. Also, an independent fuzzy model, called *dynamic range model*, is built to represent the dynamic range of the signals.

When an unknown odour has to be recognised, a fuzzy model is built to represent the shape of the unknown odour. Then this model and the dynamic range of the unknown signal are compared, respectively, with the reference shape and dynamic range models. A similarity degree between models is defined. The unknown signal is identified as being the odour with the highest similarity degree with respect to both shape and dynamic range models.

[1] E-mail: beatrice@iet.unipi.it

2. The method

For each pair sensor/odour (s, o), we build a shape model. To this aim, we consider a set of responses of the sensor to the odour in repeated experiments (*training set*). Let the time and amplitude spaces be, respectively, the time interval during which samples are taken, and the real interval to which the amplitude values of the signal belong. We choose the size of the amplitude space equal to the dynamic range of each signal in order to make the responses in the training set comparable with each other. For all responses in the training set, we uniformly partition the time and amplitude spaces into H and K intervals, respectively (possibly, $H \neq K$). The numbers H and K are application-dependent parameters. For the time space, corresponding intervals in different responses have the same lengths. For the amplitude space, corresponding intervals in different responses have lengths proportional to the pertinent signal dynamics. Then we consider complete and consistent fuzzy partitions (comprising triangular fuzzy sets) on both spaces (see Figure 1). Linguistic labels are associated with the fuzzy sets. In this way, for each response in the training set, we can associate an activation value $a_{h,k}$ with each rectangle $R_{h,k}$ determined by the Cartesian product of any two triangular fuzzy sets T_h and S_k on the time and amplitude spaces, respectively. This activation value is calculated as a function of the number and the position of the signal samples with respect to the fuzzy set modal values (a fuzzy set modal value is the point at which the membership degree is 1) as follows:

$$a_{h,k} = \frac{\sum_{i=1}^{P} \mu_{T_h}(t_i) \cdot \mu_{S_k}(s(t_i))}{\sum_{i=1}^{P} \mu_{T_h}(t_i)},$$

where P is the number of signal samples contained in $R_{h,k}$, t_i is the i-*th* time sample, $s(t_i)$ is the signal value at t_i, μ_{T_h} and μ_{S_k} are the triangular membership functions of T_h ans S_k, respectively. For each triangular fuzzy set T_h, we define a discrete fuzzy set

$$Y_h = \{a_{h,1}/S_1 + \dots + a_{h,K+1}/S_{K+1}\}$$

in the space of the labels S_k, the activation value $a_{h,k}$ being the membership degree $\mu_{Y_h}(S_k)$ of S_k to Y_h. Here, the '+' sign stands for union. The fuzzy representation of a single response of sensor s to odour o is the classical set $L_{s,o} = \{Y_1, \dots, Y_{H+1}\}$.

Then, the shape model, which represents the pair sensor/odour, is expressed in terms of the average activations of corresponding rectangles in the time/amplitude spaces of the signals in the training set.

Of course, different rectangles may have different roles in modelling a sensor response. Typically, some rectangles may discriminate between odours better than others. Weights are associated with rectangles using the Analysis of Variance (ANOVA).

Rectangle weights can also be used to associate different importance weights with sensors (e.g., a sensor weight may be proportional to the number of rectangles whose weight is greater than an appropriate threshold).

In this way, an unknown signal is first transformed into a fuzzy model using the same time/amplitude fuzzy partition as the reference shape models. Then, similarity degrees between this model and the reference shape models are calculated taking the rectangle weights into account.

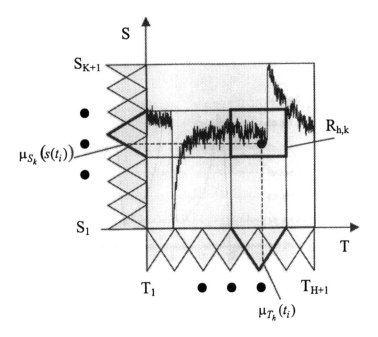

Figure 1. Fuzzy partition of the time and amplitude spaces.

As already mentioned, we also build a fuzzy model of the dynamic range of the sensor responses in the training set. This allows us to assess how similar the dynamics of the unknown signal and the dynamic ranges of the typical responses are.

When recognizing an unknown odour, for each sensor, the results of the comparison with the shape and the dynamic range models are appropriately combined. Then, a weighted sum of the responses of multiple sensors is calculated.

3. Experiments and results

The electronic nose has been applied to classify several olive oils. Here we describe a small scale application of the e-nose to the classification of 3 olive oils, namely, Lupi 100%, Borges 100%, and Parms 100%. Measurements on olive oil samples were gathered for 12 days over a period of 16 days. The sensors were exposed to odour samples and the relative sensor resistance was used for classification.

The protocol for the experiments consisted of three phases, namely sensor stabilization, absorption, and desorption of the odorant. For each experiment, the absorption and desorption phases were repeated 4 times. The sampling frequency was 1 Hz. The samples in the three phases were 30, 10 and 40, respectively, for a total of 230 samples per experiment (see Figure 2). The experiments were repeated 12 times for each oil (one repeat per day), for a total of 36 experiments.

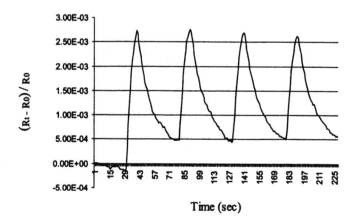

Figure 2. A sensor response in 4 repeated experiments.

The testing was performed using 6-fold cross-validation. The initial data set was segmented to provide a training set and a test set of 30 and 6 patterns, respectively. While the training set was generated by taking 10 patterns of each olive oil, the test set was created by taking the remaining 2 patterns of each olive oil. Each pattern consists of 8 sensor responses composed of 230 samples. The patterns composing the training and test sets were rotated so that in each fold we had a unique training and test set.

The recognition rates for the training and test sets were 94.33 and 72.22, respectively.

4. CP sensors

4.1. Experimental chemicals

Pyrrole from Merck was distilled and stored under nitrogen prior to use. p-Toluenesulfonate (PTS), 4-hydroxybenzenesulfonic acid (HBS), 4,5-dihydroxy-1,3-benzenedisulfonic acid (Tiron), 4-sulfobenzoic acid (SB), methanesulfonic acid (MS), 3-nitrobenzenesulfonic acid (NBS), 1,5-napthalenedisulfonic acid (NDS) and Polysodium-4-styrenesulfonate (PSS) were obtained from Aldrich Chemical Co. and used as received.

4.2. Sensor Substrate

The sensing substrate consisted of a 5 mm square silicon chip, with four parallel gold electrodes (dimensions; 15, 5, 5 and 15 μm wide with 5 μm spacing). The gold electrodes were 0.2 μm thick, and the exposed length of the electrodes was defined by a 1 μm high silicon window and was 500 μm. The chip was mounted onto PC board and wire bonded to make electrical connection. The bond wires were protected with epoxy (Epon 825/Jeffamine D230, with fused silica added to control viscosity). After curing the epoxy, the chips were vapour primed with HMDS-saturated nitrogen for two minutes. This was found to encourage lateral polymer growth across the surface of the chip, in a way similar to that reported by Nishizawa et. al. [4].

4.3. Polymer Growth and Sensing Responses

Polypyrrole (PPy) films doped with the different dopants anions were electrochemically deposited onto the gold electrodes of the sensing substrate. Deposition was carried out by applying a pulsed current (density of 1-10 mA/cm^2) and monitoring the resistance of the film as it grew (Figure 3 illustrates the process for the growth of a pTS doped PPy film). The potentiostat employed was fabricated in house and was controlled and monitored by a MacLab (ADI) system running chart (ADI) software. All films were deposited from deoxygenated aqueous solutions of pyrrole (0.1 M) and the appropriate dopant ion salt (0.1 M). The technique gave the operator control over the final resistances attained for the PPy film doped with the different anions, the optimised values for each being listed in Table 1. After deposition, the sensors were stored in a sealed container until required.

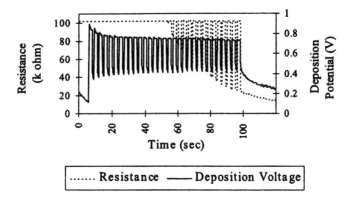

Figure 3. Deposition of pTS doped PPy using the pulsed deposition technique.

Table 1. Final resistance values for the PPy sensors doped with the different anions.

Dopant Anion	Final Growth resistance (kΩ)
pTs	4
Tiron	5
SB	0.15
NDS	0.07
NBS	4
HBS	5
MS	100
PSS	2.5

The responses of the sensors to vapours of interest were made using a 4-terminal resistance measurement, a 1 khz current and a low voltage. Its has been shown earlier that this combination gave the highest sensitivity and reproducibility [5], and eliminated the problems associated with self-heating of the polymer sensor.

5. Conclusions

In this paper a fuzzy classification method of olfactory signals has been described. An application of this method to olive oil classification using an electronic nose based on a conducting polymer sensor array has been presented. The classifier achieves 94.33 and 72.22 recognition rates on the training and test sets, respectively.

References

[1] Gardner J W and Bartlett P N 1994 *Sensors and Actuators B* **18–19** 211–20

[2] Nagle H T, Schiffman S S and Gutierrez-Osuna R 1998 IEEE Spectrum **35** 22–34

[3] Lazzerini B, Maggiore A and Marcelloni F 1998 *Electronics Letters* **34** 2229–31

[4] Nishizawa M, Shibuya M, Sawaguchi T, Matsue T and Uchida I 1991 *J. Phys. Chem.* **95** 9042

[5] Partridge A C, Harris P and Andrews M K 1996 *Analyst* **121** 1349–53

Section 3: Data Processing
Paper presented at the Seventh International Symposium on Olfaction and Electronic Noses, July 2000

133

Optimisation of an Electronic Nose for Glycerine Discrimination using the Taguchi Principle

W.G. Hansen, S.C.C. Wiedemann, H. van der Bol, V.A.L. Wortel

Uniqema, P.O. Box 2, 2800 AA Gouda, The Netherlands

This paper describes a practical methodology based on the Taguchi principle. The objective is to optimise the settings of the electronic nose parameters, in order to improve both stability and sensitivity of the discriminating sensors. When applied to glycerine, none of the selected parameters improved significantly the stability of the 5 key sensors ("SY" type). Conversely, increase of headspace temperature was found to enhance their sensitivity.

1. Introduction

Electronic Noses (EN) initially attracted the attention of the food, chemical and personal care industries for their potential benefits in routine odour analysis. However, they have so far not lived up to expectations, and their implementation in quality control laboratories of the first generation of instruments has remained limited.

The development of industrial applications has mainly been limited by the lack of reproducibility of the sensor signals [1, 2, 3]. Sensitivity of the sensors to changes in temperature and humidity has now been acknowledged [3, 4] and manufacturers have improved their stability against these external parameters, for example by temperature control of all sensor chambers. In addition, synthetic carrier gas is often preferred to ambient air to minimise the impact of any humidity fluctuations.

Sensor performance is also affected by both reversible and irreversible degradation of the surface. A tremendous effort has been made to develop more stable sensor systems [3, 5, 6], which are being rapidly integrated into commercial instruments. However, regardless of the technology, "sensor drifting" has not been completely eliminated and so remains a real challenge for quality control applications. Sensor signals have to be processed using mathematical algorithms, in order to generate reliable data. In the last five years, different compensation methods have been proposed [7, 8, 9]. They rely on a long model learning phase and are difficult to apply in practice. Very few simple procedures have been reported, and all are based on standard chemicals. Since real samples generate a wide variety of volatile organic compounds at low level, a reference (sample generating a) gas representative of the headspace can often not be found.

Optimisation of the multivariate tools has often dominated the early phases of method development, as with other secondary methods. Most of the publications have focused on improvement of the discriminative power of the model [10, 11, 12] by simply increasing the complexity degree. Proliferation of algorithms may lead to the belief that difficulties inherent in data produced by the EN have been solved. But "drifting" of the sensor responses mentioned earlier soon or later gives erroneous results. In addition, models developed for EN analysis can rarely be evaluated properly. As for any analytical technique, experimental parameters should be carefully optimised and the procedure should be properly validated. Because of the instability issue, evaluation work should of course address short- and long-term reproducibility of the EN analyses.

This paper proposes a practical methodology for the improvement of both stability and sensitivity of the sensor responses. The instrument parameters are optimised based on the Taguchi criteria. This procedure is described for a "real" application, that is the discrimination of glycerine samples.

2. Taguchi concept

The Taguchi approach [13, 14] emphasises the use of experimental design to reduce variation and move quality characteristics closer to the target. It seeks to reduce product or process variation by identifying levels of easy-to-control factors that minimise the effects that hard-to-control factors have on quality characteristics. When applied to the EN, all instrument parameters can be used as "control factors", while all parameters inducing instability or "drifting" of the sensor responses are classified as "noise factors". The latter are repeated at each set of instrumental conditions generated by the design.

The signal to noise ratio (S/N) measures the effect that noise factors have on the response variable. It is generally defined by the following equation:

$$\frac{S}{N} = 20 \ \log \left(\frac{Mean}{StandardDeviation} \right)$$

The strategy consists first of identifying the instrument parameters to optimise this signal to noise, thereby reducing instability of the EN. Next, "tuning factors" (factors that do not affect the S/N but affect the mean response) are adjusted to enhance the sensitivity of the sensors.

3. Experimental

3.1 *Equipment:*

The electronic nose instrument comprises:
➢ the Fox 4000 main unit (Alpha M.O.S.), including three "high performance chambers" in series and an injection port,
➢ the Air Conditioning Unit ACU 500 (Alpha M.O.S.) for gas humidification,
➢ the Autosampler PAL system (CTC Analytics AG), equipped with an oven for static headspace generation.

Each chamber contains six standard metal oxide sensors [15]. Furthermore, synthetic air (Hoek Loos) is used as carrier gas. It is characterised by a humidity level below 5 vpm and a hydrocarbon content lower than 0.1 vpm. Beside a Mass-flow regulator (set at 150 ml/min), an external pressure reducer (SMS) ensures precise regulation of the carrier gas pressure and hence the generation of a constant humidity level.

3.2 *Samples:*

18 Glycerine samples from 2 different origins have been collected within a 8-month period. They have been prepared within a single day to ensure consistency of both initial headspace and generated volatiles. Headspace vials (10 ml, Chromacol LTD) have been filled with 1 g (1.00 – 1.05 g) of sample using a Gilson pipette with positive displacement. All prepared vials were immediately closed and stored at 4°C in a dedicated fridge until analysed.

3.3 *EN analyses*

The system has been placed in a temperature-controlled room (+/- 2 °C). The conditions for sample analysis are summarised in the Table 1. In order to evaluate any possible memory effect of the sensors, the sequence of samples have been randomised.

Table 1: Initial parameters for headspace generation and injection

Parameters	Settings	Units
Incubation temperature	60	°C
Incubation time	25	min
Stirring rate	250	rpm
Syringe temperature	65	°C
Injection volume	4000	µL
Fill-up speed	1000	K (=µL /s)
Injection speed	2000	K (=µL /s)
Syringe flushing	240	sec

3.4 *Experimental design*

In order to select the best set of EN parameters, without running a large number of experiments, a half-fractional factorial design has been built. Based on earlier work [16], five important parameters from both headspace generation and injection have been included; with each factor at two levels (Table 2). The design consisting of 16 runs has been carried out using one average glycerine sample. Each experimental condition has been repeated 4 times to allow calculation of the Taguchi criteria. Thus, a total of 64 vials have been prepared and analysed. For each sensor and each set of experimental conditions, both average response and signal to noise ratio can then be calculated, resulting into two sets of separate data (each with 16 raws and 18 columns).

3.5 *Data processing*

From each response of the 18 sensors, only the maximum has been recorded. As a result, each headspace analysis is characterised by 18 variables. Detailed exploration of the resulting data set (variables organised as rows and vials as columns) has been done by Principal Component Analysis (PCA) [17, 18], using the software packages developed by Alpha M.O.S. (Version 6.1), and CAMO AS (Unscrambler®, Version 7.5). PCA is a projection method used to visualise all information contained in a data table by means of plots. The dimension of the data table is first reduced from 18 to 2, the first direction [PC 1] carrying most of the information from the table and the second one [PC 2] accounting for the largest residual variation. Samples and variables are then projected along these new co-ordinate axes. If necessary, further directions [PC 3 and PC4] may also be investigated. The "scores plot" helps to highlight similarities or differences between samples. The "loadings plot" describes the data structure in terms of variable correlation. The plots have to be superimposed to understand the true relations between samples and variables, and form the basis of the conclusions drawn in this report. Both design and evaluation of the experiments have been performed using SAS® (Version 8.0, SA Institute Inc.).

Table 2: Parameters included in the Half-fractional factorial design

Parameters	Low level	High levels	Units
Mass of the sample	1	5	g
Incubation temperature	60	70	°C
Incubation time	15	30	min
Injection volume	2	4	µL
Injection speed	1	2	K (=µL /s)

4. Results and discussion

A series of experiments has been designed according the Taguchi concept. The procedure has been applied to a study case, the discrimination of glycerine.

4.1 Discrimination of glycerine

Odour discrimination of glycerine is challenging, since this very polar matrix generates only very low level (below ppm) of volatiles. In addition, when produced from natural raw material, the headspace is well known to be complex and its composition may even change with the season. EN analysis of such product therefore requires the sensors to be characterised by high sensitivity, selectivity, but also high reproducibility.

In order to find out the most discriminative sensors for odour assessment of glycerine, the samples of two different origins have been analysed in triplicate. The resulting data set (54 raws and 18 columns) has been projected by PCA (Figure 1). The two types of samples (marked "C" and "E") are separated from each other along the second direction (PC2, 6% of the variance). However, the very small inter-distance confirms the similarity of the volatile compounds generated by the two groups. On the other hand, each class remains diffuse (along PC1, 88% of the variance), indicating also some variation of the headspace composition during the 8 months of collection.

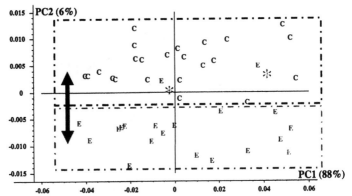

Figure 1: Discrimination of Glycerine; Projection of the samples along PC1 and PC2

The corresponding " loading plot" (Figure 2 below) is used to study the contribution of the 18 sensors to this discrimination. Most of the " SY" sensors (both Tungsten and Tin oxide) show significant y- (absolute) values and hence account for most of the discrimination. Therefore, the EN operating conditions have been optimised to enhance responses of these 5 key sensors.

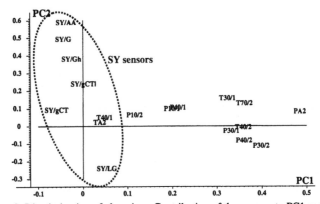

Figure 2: Discrimination of glycerine - Contribution of the sensors to PC1 and PC2

4.2 *Optimisation of the Signal to Noise ratio*

The Taguchi analysis strategy starts by identifying factors that affect the S/N ratio, in order to select optimal levels of these factors.

Figure 3 gives the projection of the experimental points collected for the 16 conditions tested; to illustrate this, each point has been marked by the corresponding sample mass. The Principal Components 1 and 2 (PC1 and PC2) account together for 86% of the total variance. When reviewing the 5 factors included, only the influence of the injection volume was revealed. Hence, all experiments run with a 2 ml injection volume are found within the top-left quadrant, while those carried out with 4 ml are grouped within the bottom-right quadrant.

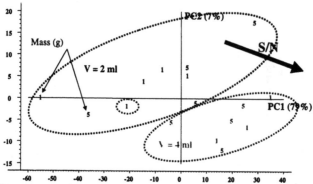

Figure 3: Analysis of S/N data- Projection of objects along PC1 and PC2

The corresponding "loadings plot" (Figure 4) gives the contribution of the sensors to the main principal components. Most "Plate" and "Tubular" sensors exhibit high x-values (PC1); thus, a high injection volume (4 ml) enhances the signal to noise ratio corresponding to these sensors; in other words, stability of the signals recorded for these sensors could be improved by increasing the injection volume.

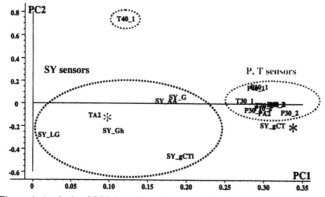

Figure 4: Analysis of S/N data - Contribution of the sensors to PC1 and PC2

Conversely, the 5 "SY" sensors of interest are not contributing significantly to the main direction (PC1), since they are all characterised by much smaller x- values. Investigation of subsequent principal components (e.g. PC3) did not show any significant effect of the experimental parameters on the signal to noise ratio obtained for the relevant sensors. The conclusions drawn based on PCA have been confirmed by Analysis Of Variance (Anova). Therefore, for this particular example, none of the factors included in the design could really help to improve the stability of the discrimination.

4.3 *Optimisation of Average response*

In a second phase, "tuning factors" affecting the average responses of the sensors of interest are identified and their levels are set to achieve the best sensitivity.

Inspection of the "loadings" corresponding to the first and second directions (PC1 and PC2) shows that the 5 "SY" sensors of interest lie together close to the centre position of the plot, and therefore make no real contribution to these principal components. Conversely, these gas sensors are lying along PC3 (Figure 5), further from the origin and are hence significantly contributing to the variance modelled by this principal component.

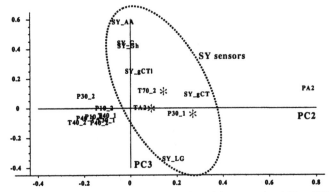

Figure 5: Analysis of mean data – Contribution of the sensors to PC2 and PC3

Figure 6 below shows the corresponding "scores plot". The experiments run with a headspace generation temperature of 60°C lie on the top half, clearly separated from those using a temperature of 70°C and situated on the bottom half. Thus, a 10°C increase of the temperature leads to significantly higher responses of the SY sensors, independently of the other parameters included in the design. Again, this conclusion was confirmed by Anova. Conversely, both methods show that the headspace temperature factor does not affect the signals produced by the other sensors.

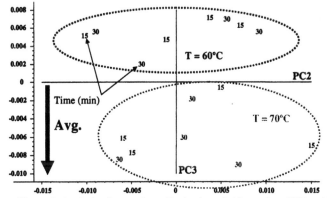

Figure 6: Analysis of mean data – Projection of objects along PC2 and PC3

5. Conclusion

The EN parameters have been optimised according to the Taguchi approach in order to improve both reproducibility and sensitivity of the sensor responses. This practical methodology has been applied to the discrimination of "real" glycerine samples.

Based on the above discussion, the following conclusions have been drawn:

➢ Out of the 18 metal oxide sensors, only 5 "SY" sensors (both tin and tungsten oxides) contribute significantly to the discrimination of the glycerine samples.

➢ None of the selected parameters improved significantly the reproducibility of the discriminating sensors. Conversely, a 10°C increase of the headspace temperature was found to enhance the selectivity.

Re-analysing the two classes of glycerine samples under the recommended conditions could validate this last point. The Taguchi approach could also have been directly applied to a parameter measuring the discrimination, keeping the number of calibration samples to a minimum.

References

[1]   Hidentsugu A, Shigehiko S and Takahashi Y 1989 Anal. Chim. Acta 225 89
[2]   Gopel W, Jones TA, Kleitz M, Lundstrom I, Seiyama T 1992 Chemical and Biochemical Sensors (Weinheim: VCH) 1-28
[3]   Zubritsky E 2000 J. Anal. Chem. 421
[4]   De Wit M, Vanneste E, Geise HJ, Nagels LJ 1998 Sensors and Actuators B 50 164
[5]   Schoneberg U, Dura HG, Hosticka, BJ, and Mokwa, W 1991 Int. Conf. Solid State Sensors and Actuators San Francisco CA
[6]   Dobos K, Strotman R and Zimmer G 1983 Sensors and Actuators 4 593
[7]   Holmberg M, Winquist F, Lundstrom I, Davide F, DiNatale C and D'Amico A 1996 Sensors and Actuators B 35 528
[8]   Holmberg M, Davide F, DiNatale C, D'Amico A, Winquist and F, Lundstrom I 1997 Sensors and Actuators B 42 185
[9]   Hugen, J-E, Tomic O and Kvaal K 2000 Anal. Chim. Acta 407 23
[10]   Sundgren H, Lunstrom and Vollmer H 1992 Sensors and Actuators 9 127
[11]   Sutter JM and Jurs PC 1997 J. Anal. Chem. 69 856
[12]   Spiegelman CH, Bennett JF, Vannucci M, McShane MJ and Cote GL 2000 J. Anal. Chem. 72 135
[13]   Kackar RN 1985 J. Quality Technology 17 176
[14]   Genter B 1987 Quality progress 44
[15]   Alpha M.O.S. 1987 Operating Manual of the Fox EN 18
[16]   Hansen WG and SCC Wiedemann 1999 EN and Sensor array based systems (Lancaster: Techtonics publishing Co) 131
[17]   Meglen RR 1991 Journal of Chemometrics 5 163
[18]   Wold S, Esbensen K and Geladi P 1987 Chemometrics and Intelligent Laboratory Systems 37

Acknowledgements

The authors gratefully acknowledge the contribution of Henkjan Abbes.

Section 3: Data Processing
Paper presented at the Seventh International Symposium on Olfaction and Electronic Noses, July 2000

141

Choice of a Suitable E-Nose Output Variable for the Continuous Monitoring of an Odour in the Environment

Nicolas, J., Romain, A.C, Monticelli, D., Maternova, J. and André, Ph.
Fondation Universitaire Luxembourgeoise
Department "Environmental Monitoring"
Avenue de Longwy, 185
B 6700 ARLON -(BELGIUM)

1.   Introduction

Among the potential application areas for electronic noses, the monitoring of our environment constitutes a real challenge. The rigorous experimental conditions which apply in the laboratory are no longer usable in the field. Particularly, the input odour can't be considered as a well-defined variation of the gaseous ambience with respect to clean air. "In the field of environmental monitoring, the background is an ever-changing chemical mixture against which we want to detect the rise of a particular odour - although the exact profile of that rise is both unknown and variable" (Gardner and Bartlett, 1999).

There is thus a need to develop instruments able to monitor continuously the odours emerging from factories, from sewage farms or from landfills sites, in order to assess the importance of public nuisance, or to control in real time the odour abatement techniques.

The department "Environmental Monitoring" at F.U.L. (Arlon, Belgium) is active in the adaptation of the electronic nose principle to discriminate some environmental odours, if possible directly in the field, and to monitor them continuously.
Research aims at improving the portability and the user-friendliness of the instrument, by trying to supply the information actually requested by the final user in the field, i.e. either the scientific operator, or the supervisor of the odorous process (Romain, Nicolas et al., 1997, 1999, 2000).

The present paper goes in the same direction by testing some practical solutions to reach those objectives.
Three main questions are put.
−   How to simplify the apparatus, in order to lower its price and to make it more easy to use in the field?
−   Which is the function most adapted to classify the various sources, and which allows to make a decision in real time?
−   What kind of signal may be used to monitor continuously the odour "intensity"?

2.   Material and methods

An array of tin-oxide sensors (either TGS from Figaro Company or multisensors from Microsens) is used in different configurations. The ambient air is either sampled around environmental sources (landfill, urban waste composting facilities, ...) and transferred to the laboratory system, or directly transferred to the sensor chamber in the mobile detector, usable directly in the field.

Various experiments, made under various conditions, led to the formulation of some practical advice concerning the design and the use of a field e-nose.

The majority of the results come however from a case study related to the odour generated by a landfill. Two kinds of odours are perceived by the neighbouring population : either the one of the fresh refuse (esters, sulphur organic compounds, solvents, ...), or the one of the biogas generated by the decomposition of the organic matter under anaerobic conditions (trace elements, such as H_2S, NH_3 and some VOC's in a mixture essentially composed of odourless compounds : methane and carbon dioxide).

Measurements are chiefly made with a mobile electronic nose constituted of 6 TGS sensors in an aluminium box (100 cm^3). A constant gas flow rate of 150 ml/min is provided by a small pump, and the system operates by a series of cycles alternating 5 minutes of pure air (sampled in a Tedlar bag from an air generator in the lab, prior to moving in the field) and 5 minutes of odorous gas transferred directly into the sensor chamber. The whole system is powered by a 12 V battery.

The operator performs measurements at some different locations on the landfill area : either in the vicinity of fresh waste, sometimes when the trucks pour out the refuse, sometimes when the waste is at rest, or at various distances from a landfill gas extraction well. At each location, he notes his feeling of odour intensity on a 4 level scale. A total of 141 observations were carried out with that procedure : 69 around "fresh waste" (including 21 observations with intensity 0), and 72 around "biogas" (including 24 zero-intensity observations).

The 6 sensor signals are recorded in a data logger and they are off-line processed by Statistica or Matlab procedures, but the calibrated model is also tested for further on-line validation in the field.

Indeed, the final goal, corresponding to the wish of the landfill manager, should be to use continuously the calibrated model in the field in order to control the atomisation of the bad odour neutralisation product.

3.   Results and discussion

Simplification of the apparatus towards a mobile and user-friendly detector.

- The test of various pre-processing data algorithms pointed out that the use of pure reference air could be avoided (i.e. the use of gas bottle or high performance filters), as long as the sensors are allowed to periodically regenerate in the presence of ambient air (Romain et al., 2000; Nicolas et al., 2000). That conclusion is drawn both for the "mobile detector" and for the laboratory system : neglecting the reference to the base line does not affect the classification of observations per source of odour.
- For the mobile detector, two types of operation are compared : the sensor array in static contact with ambient air and the use of a controlled gas flow system to transfer the odour from the source to a sensor chamber. First results show that, although the second solution gives more accurate recognition results, the static operation could be sufficient for the on-line odour monitoring (Nicolas et al., 2000). However, the influence of air movement on the heated sensors sometimes disturbs the recognition, and a sensor chamber, or at least a shelter, seems always more adequate than a true open system.

- The control of the temperature and the humidity of the gas and the thermo-regulation of the sensor chamber don't seem essential, even for outdoor operation. Some arguments may be put forward (Romain et al., 2000) :
    - the fact that TGS sensors are heated maintains the working ambience in the sensor chamber inside an appropriate temperature and humidity range;
    - the interference of ambient variables doesn't affect dramatically the odour recognition, as long as many different humidity and temperature conditions are included in the learning phase for a given odour;
    - ...

Choice of a function adapted to classify the various sources so as to be able to make a decision in real time.

Beyond the graphical appearance of classification results, showing nice ellipses around well separated groups, the end user would like to have at his disposal a simple function allowing him to clearly evaluate the membership to a given group, in order to make a decision.

In this spirit, non-supervised analyses, such as PCA, provide basically a performance evaluation of the system during the development phase, but they do not really represent learning tools aiming at the calibration of a recognition model.
On the contrary, supervised analyses, such as Discriminant Analysis (DA), or some Neural Networks algorithms, should not be regarded as evaluation tools. The first reason is that they always give satisfactory results (in particular, a total of 50 % of cases correctly classified seems a rather good result, but it is simply normal : such result is already reached with groups randomly created). Secondly, the model that they build is too specific to the operating conditions : it is valid for a given sensor array, a given gas flow rate and a given test protocol.
Now, these fixed conditions are prevailing during the final utilisation phase of the apparatus in the field. There, the e-nose is not any more in the development phase, and it must provide a function of the type "yes or no", based on a preliminary training.
The various experiments conducted until now in the field show that the classification functions provided by the DA procedures are quite appropriate to make a reliable recognition in real time, when the system is developed. Those functions are linear combinations of signals, providing as many classification scores as identified groups. A particular case is assigned to the group for which it has the highest classification score. Their use is very simple, very convenient, and leads to unambiguous classification results.
Alternatively, artificial neural network with the Radial Basis Function (RBF) architecture may be used to handle also some "unknown" class (Horner, 1999).

Choice of the useful signal that the e-nose system must provide for the continuous monitoring of the odour intensity.

As long as only the classification of odour sources is concerned, the models calibrated during the learning phase, either by statistical methods (PCA, DA, ...), or by Artificial Neural Networks, could be included in the mobile detector for the on-line identification of unknown odours. When the identification is achieved, the estimation of the odour intensity by the system can be done by appropriate techniques. Whatever the final application of the monitoring may be -either assessing the nuisance or controlling an abatement system- the useful signal should be related to the monitored odour, and shouldn't be influenced by disturbing odours or ambient parameters.

For more than ten years, several techniques are proposed in the literature (Göpel et al, 1989). In short, they use either the signal of one sensor element, or a weighted signal of all elements, obtained by a suitable regression technique. Are those proposed techniques suited for the on-line monitoring of environmental odours?

Using one of the sensor elements, preferably that with the highest sensitivity towards the identified substance, is a rather easy solution. On the landfill area, we chose to monitor the signal of TGS2610 (more sensitive to a wide variety of combustible gas). The Kendall correlation coefficient between the TGS2610 resistance and the odour intensity class, as assessed by the field operator, is -0.87 and shows a rather strong relationship between the two variables. The model obtained by linear regression allows to predict the intensity class from the recorded sensor signal. The measured intensity class is correctly predicted in 65 % of the 141 observations made on the landfill area.

However, the TGS2610 sensor is sensitive to both sources (fresh waste and biogas), and probably also to many other ones, so, its signal cannot be used to detect the rise of a particular odour among other ones. The procedure will thus always include two steps : a first identification of the odour by a classification technique, followed by the monitoring of the intensity of the global odour.

But the sensor varies also with ambient air temperature and humidity, and that is more awkward, since that kind of variation is unpredictable.

Such solution is thus applicable for "pure" substance, in the laboratory, but is not really suited to field applications, where background odours and variable ambient conditions affect the sensor response.

However, in the absence of other odours, the sole influence of ambient temperature and humidity could be taken into account (if they are measured) by a kind of "expert system". For example, the following rule can be applied :

If (the sensor resistance drops) AND (the humidity is stable or decreases) AND (the temperature is stable or decreases) THEN (there is probably an "odour event") OTHERWISE (any conclusion can't be drawn from the sensor resistance variation).

In this case, the system is able to detect the emergence of the odour only when the temperature and the humidity don't influence the sensor response. Such rule was experimented in Spring 1998 around the settling pond of a sugar factory and pointed out some "odour events", validated by the observation (Nicolas et al, 2000).

Anyway, using a weighted signal of all elements is more accurate and more "selective" to the monitored odour, i.e. minimises the cross-sensitivity (Hierold and Müller, 1989, Gall and Müller, 1989, Horner and Hierold, 1990).

If the classification model was calibrated by a supervised method, such as Discriminant Analysis (DA), the classification functions supplied as a standard result of such method can be used as "odour signal". Though they show very bad correlation with the measured intensity on the landfill, their selectivity to a given odour can be exploited for the continuous monitoring. Figure 1 shows such result : when moving away from the landfill gas extraction well, the classification function of the biogas decreases, when the one of fresh waste remains stable.

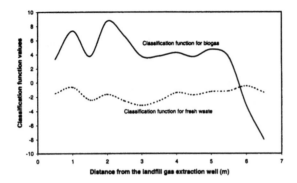

Figure 1 : Evolution of DA classification functions when moving away from the source

Different regression techniques may also be applied to predict the intensity of the odour generated on the landfill area. Multilinear Regression (MLR) on the 6 original measured sensor signals (resistances) provides a rather good model, which predicts an intensity value in agreement with the measured one in 67 % of the cases. The resulting model, however, is a pure mathematical construction, which is convenient to predict intensity values inside the training sample, but which is less adapted to the prediction of new data.

Using the results of an unsupervised classification method, such as the factors supplied by a Principal Component Analysis (PCA), has a good chance to produce a more physical model, making more "sense" from a physical standpoint (Wise and Gallagher, 1998).
Indeed, the principal component regression (PCR) includes in the model the first principal component, which is already well correlated with the odour intensity, and the second one, which separates "biogas" from "fresh waste". Including the third one in the regression provides a model which predicts the measured intensity in 69 % of the cases. Of course, the model converges towards the MLR one when the 3 remaining principal components are added. As this model MLR is worse than the model based on 3 principal components, it seems that some of the initial variables were not relevant for the prediction of the odour intensity.

Finally, Partial Least Squares regression (PLS), which captures the greatest amount of variance, like PCA, and also achieve correlation with the predictor variable (here the intensity), like MLR, will probably provide the most adapted model for the intensity prediction.
Indeed, testing PLS regression on the 141 observations on the landfill shows that the model provides 71 % of intensity prevision in agreement with the measured one. That is a very good result, knowing that, in the 29 % left, it remains probably a lot of errors of estimation due to the operator in the field. Moreover, like PCA, the PLS provides the classification of observations in two groups. Consequently, it should be used as sole tool, both to identify the source of the odour and to predicting its intensity.

References

Gall, M. and Müller, R. 1989 *Investigation of gas mixtures with different MOS Gas sensors with regard to pattern recognition.* - Sensors and Actuators, **17**, 583-586

Gardner, J.W. and Bertlett, P.N. 1999 *Electronic Noses - Principles and applications* - Oxford University Press - Oxford - 245 p.

Göpel, W., Hesse, J. and Zemel, J.N. (editors) 1989 *Sensors - A comprehensive survey (Volume I)* - VCH Verlagsgesellschaft mbH - Weinheim - 641 p.

Hierold, Chr. and Müller, R. 1989 *Quantitative analysis of gas mixtures with non-selective gas sensors* - Sensors and Actuators, **17**, 587-592

Horner, G. and Hierold, Chr. 1990 *Gas Analysis by Partial Model Building* - Sensors and Actuators B, **2**, 173-184

Horner, G. 1999 *Qualitative and Quantitative Evaluation Methods for Sensorarrays* - Proceedings of ISOEN99 - Tübingen - September 20-22, 1999, 171-174

Nicolas, J., Romain, A.C., Wiertz, V., Maternova, J. and André, Ph. 1999 *Utilisation élémentaire d'un réseau de capteurs "SnO2" pour la reconnaissance d'odeurs environnementales* - Eurodeur99, Paris, 15-18 June 1999.

Nicolas, J., Romain, A.C., Wiertz, V., Maternova, J. and André, Ph. 2000 *Using the classification model of an electronic nose to assign unknown malodours to environmental sources and to monitor them continuously.* - Accepted for publication in Sensors and Actuators B.

Romain, A.C., Nicolas, J. and André, Ph. 1997 *In situ measurement of olfactive pollution with inorganic semiconductors : limitations due to the influence of humidity and temperature* - Seminars in Food analysis, 2 (1997) 283-296.

Romain, A.C., Nicolas, J., Wiertz, V., Maternova, J. and André, Ph. 2000 *Use of a simple tin oxide sensor array to identify five malodours collected in the field* - Sensors and actuators B, **62,** 73-79

Wise, B.M. and Gallagher, N.B. 1998 *PLS_Toolbox 2.0 for use with MATLAB^{TM}* - Eigen Research, Inc. - Manson - pp.320.

Section 3: Data Processing
Paper presented at the Seventh International Symposium on Olfaction and Electronic Noses, July 2000

147

Drift Reduction for Metal-Oxide Sensor Arrays using Canonical Correlation Regression and Partial Least Squares

R Gutierrez-Osuna

Computer Science Department, Wright State University, Dayton, OH 45435, USA

Abstract. The transient response of metal-oxide sensors exposed to mild odours can be oftentimes highly correlated with the behaviour of the array during the preceding wash and reference cycles. Since wash/reference gases are virtually constant overtime, variations in their transient response can be used to estimate the amount of sensor drift present in each experiment. We perform canonical correlation analysis and partial least squares to find a subset of "latent variables" that summarize the linear dependencies between odour and wash/reference responses. Ordinary least squares regression is then used to subtract these "latent variables" from the odour response. Experimental results on an odour database of four cooking spices, collected on a 10-sensor array over a period of three months, show significant improvements in predictive accuracy.

1. Introduction

Besides selectivity and sensitivity constraints, sensor drift constitutes the main limitation of current gas sensor array instruments [1]. In order to improve baseline stability, it is customary to expose the sensor array to a sequence of wash and reference gases prior to sampling a target odour. Since wash/reference (W/R) gases can be assumed to remain stable over time, variations in sensor response during these stages may be utilized, at no extra cost, as measures of the drift associated with each experiment and, therefore, compensated for in software.

The instrument employed in this study is an array of ten commercial metal-oxide sensors. During operation, the array is initially washed for ten seconds with room air bubbled through a two-per-cent n-butanol dilution in order to flush residues from previous odour samples. This wash stage is followed by a three-minute reference stage with charcoal-filtered air, which provides a baseline for the subsequent ninety-second odour stage, in which the array is finally exposed to the target odour. Figure 1 illustrates a typical sensor transient response obtained with this procedure.

To preserve information from the sensor dynamics, we apply the windowed time slicing (WTS) technique [2, 3] to compress each exponential-shaped transient down to four weighted integrals, as shown in the lower portion of Figure 1. With ten sensors, this compression technique yields 40-dimensional feature vectors x_W, x_R and y for the wash, reference and odour transients, respectively. x_W and x_R are subsequently combined to form a regression vector x with 80 dimensions.

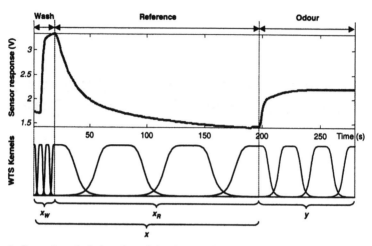

Figure 1. Extraction of window-time-slicing features from the wash, reference and odour stages

2. Drift reduction approach

Whereas variance in the W/R vector x is primarily due to drift, variance in y is a result of the combined effect of drift and the odour-specific interaction with the sensing material. Therefore, we seek to remove variance in y that can be explained by (correlated with) x while preserving odour-related variance. We propose a drift reduction algorithm that consists of three steps:

1. Find linear projections $\tilde{x} = Ax$ and $\tilde{y} = By$ that are maximally correlated:
$$\{A, B\} = \arg\max[\rho(Ax, By)] \tag{1}$$
   Vectors $\tilde{x}$ and $\tilde{y}$ are low-dimensional projections that summarize the linear dependencies between x and y.

2. Find a regression model $y_{pred} = W_{OLS}\tilde{y}$ using Ordinary Least Squares (OLS):
$$W_{OLS} = \arg\min|y - W\tilde{y}|^2 \tag{2}$$
   The OLS prediction vector y_{pred} contains the variance in the odour vector y that can be explained by $\tilde{y}$ and, indirectly, by the W/R vector x.

3. Deflate y and use the residual z as a drift-reduced odour vector for classification purposes:
$$z = y - y_{pred} = y - W_{OLS}By \tag{3}$$

To find the projection matrices A and B for the first step of the algorithm we employ Canonical Correlation Analysis (CCA) and Partial Least Squares (PLS.) CCA is a multivariate statistical technique closely related to Principal Components Analysis (PCA.) Whereas PCA finds the directions of maximum variance of each separate x- or y-space, CCA identifies directions where x and y co-vary. It can be shown [4] that the columns of the projection matrices A and B are the eigenvectors of:

$$\left(\Sigma_{xx}^{-1}\Sigma_{xy}\Sigma_{yy}^{-1}\Sigma_{yx} - \lambda I\right)a = 0$$
$$\left(\Sigma_{yy}^{-1}\Sigma_{yx}\Sigma_{xx}^{-1}\Sigma_{xy} - \lambda I\right)b = 0 \tag{4}$$

where $\{\Sigma_{xx}, \Sigma_{yy}\}$ are the "within-space" covariance matrices and $\{\Sigma_{xy}, \Sigma_{yx}\}$ are the "between-space" covariance matrices. The first pair of "latent variables" (canonical variates) $\tilde{x}_1 = a_1 x$ and $\tilde{y}_1 = b_1 y$ corresponds to the eigenvectors a_1 and b_1 associated with the largest eigenvalue of (4). This eigenvalue λ_1, identical for both equations, is the squared canonical correlation coefficient between $\tilde{x}_1$ and $\tilde{y}_1$. Subsequent canonical variates are associated with decreasing eigenvalues and are uncorrelated with previous variates. CCA will find as many pairs of canonical variates as the minimum of P and Q, the dimensionality of x and y, respectively.

Alternatively, one may use PLS to find the canonical variates (scores) $\tilde{x}_k$ and $\tilde{y}_k$. It must be noted that, whereas CCA finds all the eigenvector pairs (loadings) a_k and b_k simultaneously, PLS operates sequentially, extracting one pair of loadings/scores at a time using a NIPALS algorithm [5]. Deflation of x and y in PLS is also performed sequentially, after the extraction of each pair of loadings/scores. Furthermore, CCA maximizes the correlation coefficient between each pair $(\tilde{x}_k, \tilde{y}_k)$ while PLS takes into consideration both correlation and variance in x and y [6]. Both CCA and PLS versions of the drift-reduction algorithm have been implemented and are reported in this article.

3. Preliminary analysis

The proposed drift-reduction method is evaluated on an odour database of four spices (ginger, red pepper, cumin and cloves) containing 374 examples collected during 24 different days over a period of three months. The array consists of ten metal-oxide sensors from Capteur® (models AA14, AA20, G5, AA25, G7, LG9 and LG10) and Figaro® (models 2620, 2610 and 2611.)

We perform a preliminary analysis by feeding the entire database to the drift-reduction algorithm. The resulting correlation coefficients for each pair of latent variables $\rho(\tilde{x}_k, \tilde{y}_k)$ are shown in Figure 2. This result indicates that PLS is able to reduce correlations between x and y with fewer components than CCA, a reasonable result since PLS performs deflation sequentially, after extraction of each "latent variable." The figure also suggests that there are 10 major canonical variates (interestingly, the same as the number of sensors), although Bartlett's sequential test [4] indicates that there are 21 statistically significant variates.

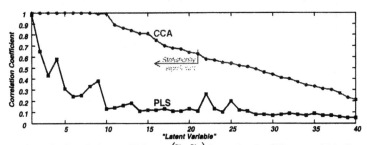

Figure 2. Correlation coefficient $\rho(\tilde{x}_k, \tilde{y}_k)$ for each pair of "latent variables"

To illustrate the performance of the algorithm, we select the first 10 "latent variables" and plot the trajectory of the four WTS features over time before and after drift reduction. Figure 3 shows these trajectories for the AA20 sensor. Examples are sorted by class –notice the four distinctive blocks— and, within each class, by date. Drift-reduced trajectories have been shifted vertically for visualization purposes. Both CCA and PLS versions of the algorithm significantly reduce the linear component of the drift. In other words, the algorithm is able to reduce drift-related variance (noise) while preserving odour-related variance (information.) It is also interesting to notice that the first WTS window is deflated almost entirely, a reasonable result since it is highly correlated with the last WTS window of the reference transient.

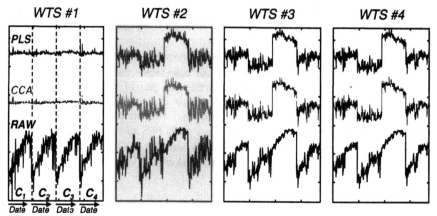

Figure 3. Trajectory of WST features before and after drift reduction

4. Analysis of predictive accuracy

A more interesting and formal assessment of performance is predictive accuracy, that is, the ability of a subsequent pattern classifier to correctly classify previously unseen test data. In the context of drift-reduction, predictive accuracy should be measured by using test examples that were collected on different days than those used for training, as illustrated in Figure 4. Formally, given a database collected over M days, we seek to classify data from day k using data from days i through j, with $i<j<k$. We denote by W (width) the number of days used for training ($W=j-i+1$) and D (distance) the number of days between the training and test days ($D=k-j$.) These two parameters, also shown in Figure 4, can significantly affect the predictive accuracy of a classifier:

- With increasing values of W, the training set incorporates data from more days, which may allow a classifier to automatically average out the drift component.
- With increasing values of D, the effects of drift accumulate on the test set, which may result in lower predictive accuracy.

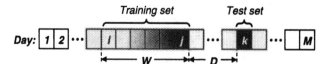

Figure 4. Illustration of W and D for predictive accuracy measurements

To analyze the effect on predictive accuracy of these two parameters along with the proposed drift-reduction technique, we use a k-Nearest-Neighbor (kNN) classifier [7]. This voting rule classifies an unlabeled example by choosing the majority class among the k closest examples in the training set. Our implementation uses the Euclidean metric, with the number of neighbors being set to one-half the average number of examples per class in the training set. To limit the "curse of dimensionality," the feature vectors y and z are pre-processed with PCA, and the largest eigenvalues containing 99% of the variance are selected. PCA eigenvectors are computed from the training set and then used to project test data. Depending on the number of training examples (a function of W,) this results in 6 to 10 principal components being retained for y and 3 to 6 principal components for z. As shown in the previous section, the drift-reduction algorithm significantly deflates the first WST feature, which explains why fewer principal components are required to total the same percentage of the variance.

It must be noted that this new analysis is more challenging than the preliminary validation shown in Figure 3, in which the entire dataset (days 1 through 24) was passed to the algorithm. In the more realistic scenario described in this section, only past training data (days i through j) and current test data (day k) are used to perform drift reduction. Intuitively, with fewer training data, fewer "latent variables" should be used to deflate the odour vector y. In fact, using ten "latent variables" yields only modest improvements in predictive accuracy compared to raw data. After some experimentation, we determined that three was a suitable number of "latent variables." We also noticed that in order to eliminate the linear component of the drift (see Figure 3) it was necessary to incorporate timing information into the drift reduction algorithm, especially for large values of D. For this reason, we decided to augment the regression vector x with the time stamp of each sample. This had not been necessary in the previous section, where the entire sequence of days was presented to the drift-reduction algorithm. With the selection of the first three "latent variables" and the addition of time stamps, we were able to obtain significant improvements in predictive accuracy, which are reported below.

The effect of distance on predictive accuracy is shown in Figure 5 for different values of the width parameter. As expected, the performance of the raw data decreases considerably with increasing distances (older training data.) After the application of CCA- or PLS-based drift reduction, predictive accuracy does not decrease with distance, an indication that the algorithm is able to compensate for drift. In fact, CCA and PLS present a slight increase in predictive accuracy with distance. This counter-intuitive result could be attributed to the fact that both training data and unlabeled test data are used to estimate the drift model $y_{pred} = W_{OLS}By$, making the slope $W_{OLS}B$ more accurate with increasing distances as a result of the time stamps. This conjecture, however, deserves further attention.

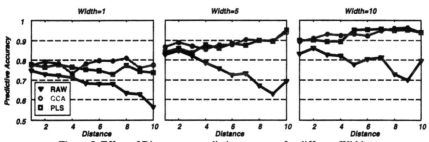

Figure 5. Effect of Distance on predictive accuracy for different Widths

The effect of width is shown in Figure 6. The three datasets yield higher predictive accuracy with increasing values of width (larger training sets.) Higher slopes in the CCA and PLS curves indicate that the drift-reduction technique allows the subsequent pattern classifier to make better use of additional training data up to a saturation point of 95% predictive accuracy, which can be clearly observed in the rightmost plot of Figure 6. No significant differences in predictive accuracy between CCA and PLS can be found in these two figures.

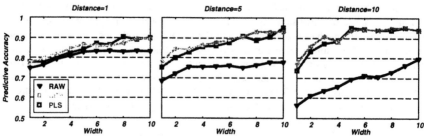

Figure 6. The effect of Width on predictive accuracy for different Distances

5. Conclusions and future work

We have proposed a drift-reduction algorithm that takes advantage of drift information contained in the pre-wash/reference stages of a typical sampling cycle. The algorithm has been shown to reduce drift-related variance in the sensors and, therefore, enhance class discrimination and predictive accuracy of a subsequent pattern classifier.

Our experience has shown that the predictive-accuracy performance of the algorithm is sensitive to the number of "latent variables," which has been manually selected in the current implementation. Automated selection of "latent variables" by cross-validation techniques constitutes the next natural step of this work. Further improvements could be obtained by augmenting the regression vector x with temperature, humidity and mass flow information, as well as transient information from appropriate calibration mixtures that could be added to the sampling cycle. Finally, the proposed algorithm is limited to linear dependencies between W/R and odour stages. Non-linear extensions of CCA [8] and PLS [9] will be required for situations where these relationships are markedly non-linear.

6. Acknowledgements

This research was supported by awards WSU/OBR 0409 and NSF/CAREER 9984426. We would like to acknowledge Alex Perera (Universitat de Barcelona) for his valuable comments and suggestions on earlier drafts of this document.

7. References

[1]   Gardner J W and Bartlett P N 1999 *Electronic Noses: Principles and Applications* (Oxford: Oxford Univ. Press)
[2]   Kermani B G 1996 *On Using Artificial Neural Networks and Genetic Algorithms to Optimize Performance of an Electronic Nose* Ph.D. Dissertation, North Carolina State University.
[3]   Gutierrez-Osuna R and Nagle H T 1999 *IEEE Trans. Syst. Man Cybern. B* 29(5), 626-632.
[4]   Dillon W R and Goldstein M 1984 Multivariate Analysis: Methods and applications (New York: Wiley)
[5]   Geladi P and Kowalski B R 1986 Analytica Chimica Acta 185 1-17.
[6]   Burnham A J et al. 1999 J Chemometrics 13 49-65
[7]   Duda R O and Hart P E 1973 *Pattern classification and Scene Analysis* (New York: Wiley)
[8]   Lai P L and Fyfe C 1999 Neural Networks 12(10), 1391-1397.
[9]   Malthouse E C et al. 1997 Computers chem. Engng 21(8) 875-890

Section 3: Data Processing
Paper presented at the Seventh International Symposium on Olfaction and Electronic Noses, July 2000

153

Sensor Array Data Processing using a 2-D Discrete Cosine Transform

B Naidoo, A D Broadhurst

School of Electrical & Electronic Engineering, University of Natal, Durban, 4041, South Africa, Fax: +27 31 260 2111, E-mail: bnaidoo@nu.ac.za

Abstract. A novel technique is presented for representation and processing of sensor-array data as a two-dimensional (2-D) image. The near optimal, computationally efficient 2-D Discrete Cosine Transform (DCT) is used to represent the 2-D signal in a decorrelated spatial frequency domain. The semantics of the transform domain representation are explored and various techniques are presented for the gainful manipulation of transform domain coefficients. A heuristic technique is developed for the selection of transform domain coefficients as input to a non-linear classifier. Several benefits are evident in the results. These include; significant dimensionality reduction and concomitant reduction in classifier size and training time, improved generalisation and improved classification performance.

1. Introduction

A low cost electronic nose based on commercially available SnO_2 sensors was developed [1,2]. Rigid sampling regimes and refined signal processing techniques were used to redress the non-ideal nature of the sensory front end. This paper deals primarily with the signal processing heuristic that was evolved in response to the initially poor performance.

The system was based on an array of Figaro Taguchi Gas Sensors (TGS). Its purpose was the classification of cheese odours.

2. The Sampling Event

Once off 'snap-shot' measurements can easily lead one astray because the sensors have difficulty reaching a meaningful steady state. It is therefore necessary to capture a time series measurement of the sensory response to a controlled stimulus. Such measurements are also able to capture temporal patterns [3] in the signal. The time series measurement and odour stimulus control schedule is referred to as a sampling event.

2.1 *Measurement.* An array of 6 different sensors is sampled simultaneously at 1Hz over 256 samples. The 6 x 256 measurement matrix is padded with zeros to make it 8 x 256 (i.e. two dummy channels are added). The fast DCT signal-processing algorithm requires this. The measurement matrix represents a single odour response pattern.

2.2 *Stimulus control schedule.* The odour stimuli are applied according to a strict schedule. See Table 1.

Table 1 : Measurement phases

Phase	Description
A	20 samples are taken with the sensor in a resting state (ie. no stimulus is applied).
B	30 samples are taken with pure (synthetic) air flowing through the sensor head.
C	50 samples are taken with flowing odourant/air mixture.
D	40 samples are taken with static odourant/air mixture trapped in the sensor head.
E	116 samples are taken with air flushing the sensor head.

Phases A and B encode the sensors' base reference and their response to a standard flowing air. This indicates the sensory status prior to the odourant stimulus. Phase C encodes the flowing odourant response transient. Phase D captures the response to a static odourant mixture, while phase E records the sensors' recovery transient. Therefore, each pattern provides a rich information source.

3. Decorrelation

Orthogonal transforms may be used to transform natural data sequences into a representation that contains little, or no redundancy. That is, the sequence is decorrelated in the transform domain; where each value (coefficient) is unrelated to the others and may be processed independently. Redundancies are removed so the area under the curve is minimised without any loss of information and a maximum amount of information is stored in the fewest number of coefficients. The transform domain representation may be regarded as an exact (lossless) but more concise reproduction of the original information. Decorrelated sequences are usually more conducive for Artificial Neural Network (ANN) [4] based classification.

Furthermore, information content is concentrated in a few major coefficients. This allows for the majority of coefficients to be removed from the data set without significant information loss. The reduction in pattern dimensionality reduces the ANN capacity requirements (number of trainable connections).

Coefficients in the transform domain represent bases (sub-patterns) in the source domain. Removal of coefficients corresponds to the removal of sub-patterns in the source domain. Selective coefficient removal can eliminate disruptive sub-patterns (eg. high frequency noise). The remaining coefficient set may be more focused and separable than the original data. Two orthogonal transforms are considered.

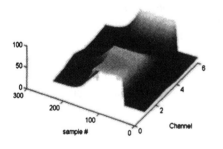
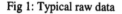

Fig 1: Typical raw data

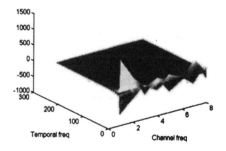

Fig 2: Typical DCT representation

3.1 *The Optimal Transform*

The Karhunen-Loêve Transform (KLT) technique, also known as Principal Component Analysis (PCA) [5] is able to completely decorrelate the sequence in the transform domain. It packs the maximum information in the fewest number of independent transform coefficients [6].

Difficulties associated with the KLT or PCA:
- There is no fast algorithm for its implementation.
- It is computationally expensive.
- It requires large amounts of data.
- It is not a fixed transform, but is dependent on the signal statistic. Therefore its performance can be adversely affected by variations in the signal statistic (eg. due to sensor replacement or drift).

3.2 *The Near Optimal Transform*

The Discrete Cosine Transform (DCT) was shown to be the best performer of the near optimal transforms. DCT performance compares closely with the KLT in terms of information distribution across its coefficients [6].

Advantages associated with DCT:
- Near optimal performance.
- Numerous fast algorithms exist for its implementation.
- It is a fixed transform that is independent of the signal statistic.

4. Coefficient Manipulations

The DCT was used to transform the 2-D channel-time domain (Figure 1) to the 2-D channel frequency - temporal frequency domain (Figure 2). Various modifications were effected in the transform domain. A new data set was saved after each modification, and a comparative study was performed on the separability of the differently transformed data sets. We began by compensating for sensor drift by zero weighting the DC temporal frequency coefficients (D1 data set). D1 pattern size was then reduced by zero weighting the small amplitude high-frequency temporal coefficients (D2 data set - see Figure 3). Subsequent data sets were generated by a coefficient selection heuristic.

4.1 *Coefficient selection heuristic*

4.1.1 *Summary.* A coefficient selection heuristic was developed to select a reduced set of coefficients from the existing patterns. The heuristic performs a statistical analysis on the pattern database and ranks the coefficients according to a linear statistical separability measure. The discrete channel frequency axis has eight frequency bins. The most highly ranked coefficient in each bin was selected. This reduced the pattern size down to eight coefficients that spanned the channel frequency axis.

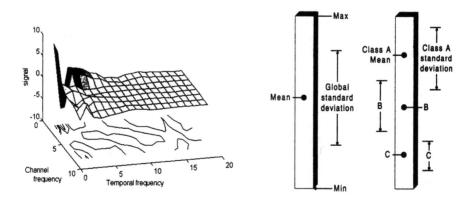

Fig 3: Typical D2 data   Fig 4: Coefficient statistics

4.1.2 Details.

Selective removal of transform domain coefficients is common in image compression techniques. For compression, those coefficients with the highest variance are retained because they encode the most information [7]. Preservation of variance is important for the eventual reconstruction of the image stream. Variance also plays an important role in classification problems where coefficients with high variance encode larger amounts of differential information across categories. The intention of coefficient selection is the improvement of separability.

A statistical analysis is performed on the entire data set in the transform domain. As indicated in Figure 4, a global mean and global standard deviation can be calculated for each coefficient. These calculations initially cover the entire data set and are not category specific. Assuming that the data is separable, a larger standard deviation indicates that the data is more spread and possibly easier to separate. That is, more room is available for the formation of decision boundaries. So the simplest ranking would be one based purely on global standard deviation or variance.

A priori findings showed that it was essential for the final set of coefficients to span the channel frequency axis. This is further illustrated in the results section. Therefore, the coefficient with the highest variance in each channel-frequency bin was selected. This gave a final selection of 8 coefficients, producing a data set that was 256 times smaller than the original. The resultant data set is called the D3A data set.

Depending on the actual signal statistic, global variance can be a misleading indicator of separability. For example, a large variance could indicate that the individual categories are far apart and easily separable, or it may be indicating the categories are spread out and overlapping.

A simple modification provides an effective remedy to the above dilemma. Calculate the individual category standard deviations (for each coefficient) as indicated in Figure 4 to the right. Then find the average (mean) category standard deviation for each coefficient. For maximum separability, we need two things:

- Category standard deviations must be small - indicating that the data are packed closely around their respective means.
- Category means must be far apart (large global standard deviation) - indicating low likelihood of overlapping categories.

Table 2: Data sets

Code	Description
D1	Transform domain representation with DC temporal-frequency components removed.
D2	Same as D1 but with low amplitude high-frequency components removed as well.
D3A	8 highest variance coefficients chosen such that the channel-frequency axis is spanned.
D3B	8 highest normalised-variance coefficients that span the channel-frequency axis.
D3C8	Similar to D3B but the coefficients need not span any axes.

Dividing the global standard deviation by the mean category standard deviation gives a far more reliable separability indicator.

Using this technique, the coefficients are ranked once again, and one coefficient is selected for each channel-frequency bin. This gives a final selection of 8 coefficients per pattern. The resultant data set is called the D3B data set. For completeness, a third data set was generated where only the top 8 coefficients were selected without having to span the channel-frequency axis. This is called the D3C8 data set.

5. Results

Comparative results for the various data sets are provided in Table 3. Since this was a comparative test, all random number generators in the signal processing software and the neural network simulator (Stuttgart neural network simulator) were seeded with identical values for each experiment. This produced consistent pattern distributions across the various test, validation and training sets. It also ensured that the neural network simulator presented the same training sequence to the various neural networks. The neural network with the best validation result was saved for each data set. Each neural network was tested with its associated test set. The results are given in Table 3.

D1 data set: This data set required large ANN sizes and long training times.

D2 data set: Classification rates only reached acceptable levels when the hidden layer size reached 500 units. This exceeds the number of training patterns and indicates either a non-representative training set, or a training set separability problem. However, we know that there is no representation problem because the same training patterns are used in the more successful data sets (after further processing).

D3A data set: Poor performance indicates that the sole use of global variance as an indicator of separability can be misleading. It could also indicate that eight coefficients are too few.

Table 3: Summary of results

	Pattern size	Classification rate	Training time	Hidden layer size
Original data	8 x 256	97.77%	17:01:22	200
D1 data	8 x 255	96.88%	23:02:17	200
D2 data	8 x 19	99.11%	10:20:42	500
D3A data	8	35.27%	00:16:00	175
D3B data	8	99.11%	00:00:36	12
D3C8 data	8	35.27%	00:03:13	40

D3B data set: Good performance shows the effectiveness of the coefficient selection heuristic. It proves that the training set was representative of the general class of signals, and it shows that eight coefficients are sufficient for successful classification. A drastic speeding-up of the training process was also noted. This is due to reduced input dimensionality, reduced hidden layer size and greatly improved separability.

D3C8 data set: Poor performance indicates that spanning of the channel-frequency axis is necessary. This also illustrates the relative importance of the chemical selectivity of the sensors (represented by the channel-frequency axis in the transform domain).

Summary: The D3B data set which was processed according to the coefficient selection heuristic produced good classification results and rapid training. It also required a much smaller classifier thus indicating improved separability.

6. Conclusion

The coefficient selection heuristic is trained on the signal statistic. Therefore its performance could degrade if the signal statistic should change (e.g. due to sensor drift).

The need to span the channel-frequency axis indicates that, for this experiment, chemical selectivity is more valuable than temporal information. The value of the temporal dimension can be improved by using a gas separation column.

The improvement in classifier performance with the careful selection of coefficients indicates that separability was improved. The fact that a small neural network (with fewer connections) was able to produce the same result as a much larger one shows us the data set used with the smaller network possessed a more focused and useful information content.

References

[1]   D.C. Levy & B. Naidoo, How machines can understand smells and tastes: Controlling your product quality with neural networks, Ch 22, pp. 199-207, in G.A. Bell & A.J. Watson (eds.), Tastes & Aromas: The chemical senses in science and industry, UNSW Press, Sydney, Australia.

[2]   B. Naidoo, D.C. Levy, G.A. Bell & D. Barnett, Food odour classification in an artificial olfactory system, International conference on engineering applications of neural networks (EANN - 1995), Finland.

[3]   M. M. Mozell, Olfactory Discrimination: Electrophysiological Spatiotemporal Basis, Science,143, 1964, pp. 1336-1337.

[4]   J. W. Gardner, E. L. Hines & H. C. Tang, Detection of vapours and odours from a multisensor array using pattern-recognition techniques, Part 2. Artificial neural networks, Sensors and Actuators B, 9 (1992), pp. 9-15.

[5]   J. W. Gardner & P. N. Bartlett, Pattern Recognition in Odour Sensing, Ch 11, pp. 161-180, in J. W. Gardner & P. N. Bartlett (eds.), Sensors & Sensory Systems for an Electronic Nose, ASI Series, Springer-Verlag, Berlin

[6]   N. Ahmed, T. Natarajan & K. R. Rao, Discrete Cosine Transform, IEEE Trans. Comput., vol. C-23, Jan. 1974, pp. 90-93.

[7]   H. C. Andrews, Multidimensional rotations in feature selection, IEEE Trans. Comput., vol. C-20, Sept. 1971, pp. 1045-1051.

Section 3: Data Processing
Paper presented at the Seventh International Symposium on Olfaction and Electronic Noses, July 2000

159

Identification of Pollutant Gases and its Concentrations with a Multisensorial Arrangement

S.Reich [(1)], R.M.Negri [(2)], A. Lamagna [(3)] and L. Dori [(4)].

(1) Escuela Ciencia y Tecnología, Universidad de San Martín, Argentina
(2) INQUIMAE, Facultad de Ciencias Exactas, Universidad de Buenos Aires. Argentina.
(3) Departamento de Física, Comisión Nacional de Energía Atómica, Argentina
(4) CNR-Istituto Lamel, Bologna, Italy

Abstract, In this work we study the performance of modeled electronic noses to elucidate several aspects of the pattern recognition scheme. We analyzed electronic noses based on two different types of sensors: tin dioxide commercial sensors and micromachined sensors developed in our laboratories tailored to identify gases of relevance in polluted atmospheres. For commercial sensors we analyze the possibilities of determining both the concentration and composition of a mixture of gases containing carbon monoxide, methane and isobutane. Several aspects concerning the identification of individual gases in the presence of noise are studied for e-noses based on the second type of sensors.

1. Introduction

We analyze the capability of a modeled electronic nose to identify gases of relevance in polluted atmospheres, mainly CO and NO_2 and its concentrations, when other interfering gases are present and under noisy conditions of detection. The first step, due to the availability of complete calibration curves for individual chemical species, was to implement a theoretical nose, composed by a few number of commercial tin dioxide sensors (Taguchi), able to identify the presence of CO, ISBU, CH_4 and EtOH. Gas concentrations range between few hundred to thousand parts per millions, which is a characteristic of Taguchi sensors. We focus in the identification of the individual gases and its concentrations when present in a mixture.

In the second part of the work, a model nose is simulated using the experimental output of 100 nm thick SnO_2 sensors deposited upon a substrate heater element made on micromachined silicon with a 200 nm Si_3N_4 membrane as physical support of the entire sensor stack [1]. The sensitivity of these sensors is in the range of a few ppm, so are highly adequate for environmental monitoring of pollutant gases. In this case, CO, NO_2, benzene, and a mixture of toluene and xylene (TX) were essayed at the ppm levels.

2. Modeled nose based on Taguchi type sensors

We have modeled the overall nose response vector to a given sample when two or more gases, for which each sensor has a non-negligible response, are simultaneously present. In a previous work, Bartlett and Gardner[2] have used a phenomenological linear model to simulate the response of semiconductor sensors to such a mixture. The concentration dependence of a sensor signal in a gas mixture of N components of concentrations $\{c_1,...., c_N\}$ was calculated as a linear combination of the sensor responses to every individual gas in a pure system. The linear response dependence assumed by Barlett and Gardner is generalized in this work.

For a sample consisting of N gases, the total concentration c of the mixture is related to the concentration of each gas c_k:

$$c = \sum_{k=1}^{N} c_k \qquad (1)$$

We define the relative proportion α_k as:

$$\alpha_k = \frac{c_k}{c} \quad ; \quad \sum_{k=1}^{N} \alpha_k = 1 \qquad (2)$$

For the particular case ($\alpha_k = 1$, $c = c_k$) the response of the sensor i, R_k^i is:

$$R_k^i = \frac{A^i(1+\varepsilon_A \Delta_A)}{1+B_k^i(1+\varepsilon_B \Delta_B)c_k^{m_k^i(1+\varepsilon_m \Delta_m)}} \qquad (3)$$

The parameters A^i, B_k^i, $m_k^i, \Delta_A, \Delta_B$ and Δ_m (that are dependent on the sensor i and the gas k) were obtained fitting the calibration curves and are given in reference [3].; ε_A, ε_B, and ε_m are random numbers in the interval [-1, 1].

For a two-gas mixture that has a given total concentration c, and with proportional fractions α_1 and α_2, the response of sensor i measured by its relative resistance is:

$$F^i(c_1,c_2) = \tilde{R}_1^i(c) + (\tilde{R}_2^i(c) - \tilde{R}_1^i(c))g(\alpha_1) \qquad (4)$$

where R_k^i corresponds to gas k with concentration c measured by sensor i and $g(\alpha)$ is any function that interpolates the modeled value such that $g(0)=1$ and $g(1)=0$. In the above formula all the R_k^i are evaluated for a given value of c and the varying parameter is the concentration fraction α_k of each component. This approach is equivalent to that of reference [2] when the extrapolation function is linear and the sensor response for each gas is also linearly dependent with concentration.

Generalizing, if a third gas is present in the mixture with fraction α_3 and such that the total mixture concentration is also c, then the relative resistivity of sensor i is assumed to be given by:

$$F^i(c_1,c_2,c_3) = \tilde{R}_1^i(c) + (\tilde{R}_2^i(c) - \tilde{R}_1^i(c))g(\alpha_1) + (\tilde{R}_3^i(c) - \tilde{R}_2^i(c))g(\alpha_1)g(\alpha_2) \qquad (5)$$

Thus, a simple recursion formula can be implemented for an arbitrary number of active gases. From the wide variety of scenarios of samples and modeled responses, we have chosen the identification of CO and its concentration when it is sampled in combination either with CH_4 or with EtOH. The modeled nose consists of four Taguchi sensors. The sample space contains CO with significant mixture of either CH_4 or EtOH. Total concentrations vary between 500 and 2500 ppm, and with α_1 covering all ranges between 0 and 1 in steps of 0.1. A linear function for $g(\alpha)$ was chosen for the sake of simplicity but of course more complex options can be chosen.

We generated about 450 samples for each mixture using the above procedure. Principal components, with 93% of the overall projection concentrated in the first two components are shown in Figure 1. The corresponding graph exhibits an overlapping region. This region corresponds to mixtures where α_1 corresponding to CO is above 0.8. The best strategy found for identification of a particular gas, its proportion in the mixture and the total concentration of the mixture is sketched in Figure 2. The response of each sensor i (F^i) is first broadcast to a network A with two hidden layer and one output unit (4-2-1) that identifies the particular mixture between the two possibilities (mixture with CH_4 or EtOH). Then, the relative resistivity (F^i) of each sensor is input to a second network B (5-5-2) together with the first output. This second network B is separately trained for each of the two combination of gases. A 20% of the generated response vectors were randomly chosen for the training set. The generalization is done with the rest (80%) of the samples.

The performance for gas identification, (network A) has of more than 95% of efficiency if the gas other than CO was present with a molar fraction grater than 0.1. In the case of the other gas predominant with more than 89%, the first network has a percentage of confidence of about 60%, consistent with a random election between the two possibilities. In spite of this, the presence of CO and its concentration is correctly identified.

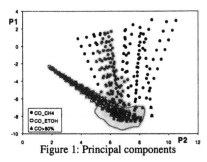

Figure 1: Principal components

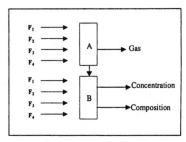

Figure 2: Pattern recognition scheme

Results for concentration and composition are shown in Figure 3 and Figure 4. In Figure 3 we plotted the percentile for the absolute values of the relative error of the adjustment between the network identification of the concentration of the mixture and the "real" ones. It indicates that 85% of the times the concentration is estimated with less than 10% of relative error. Figure 4 shows, for the relative proportional fraction α, the retrieved values versus the real ones. As it can be observed the adjustment is fairly good. The trend line and the R^2 of it are displayed in the plot making evident the adequate fitness.

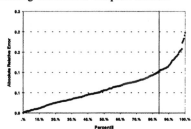

Figure 3: Assignment of concentration

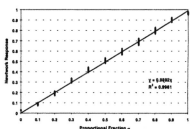

Figure 4: Assignment of gas fraction

3. Micromachined sensors

Two different groups of sensors were produced, both of them were deposited upon a substrate heater element placed on micromachined silicon and having 200 nm Si_3Ni_4 membrane as physical support of the entire sensor stack. The technique used to deposit the SnO_2 layer was the Rheotaxial Growth and Thermal Oxidation (RGTO). The thermal evaporation deposition of metallic Sn was done at a constant rate (0.5nm/s) on the substrate kept at 400 °C in vacuum. The second step consists of the metal thermal oxidation of the Sn film. In order to obtain the whole oxidation of the film, the samples were annealed at 620 °C in flux O_2 atmosphere for 32 hours. These sensors have a 100nm thick SnO_2 layer.

The film structure and morphology were studied using X ray diffraction (XRD) analysis and scanning electron microscopy (SEM) observation, respectively. The XRD analysis showed a fully oxidation of the metallic tin first layer. The SnO_2 films grown by the RGTO technique produces polycrystalline films with high surface area and with grains connected by necks. The performance of the sensing layers were evaluated by using sequentially gas mixtures containing C_6H_6 (1 and 5ppm), CO (10 and 30ppm), NO_2 (0.1 and 0.5ppm) and with TX (1 and 5ppm) in synthetic air with a relative humidity of 30 %. This gas sequence was periodically repeated to monitor the sensor signal stability and reproducibility. During the functional test, the SnO_2 layer was heated at 400°C by a Pt heater resistor integrated on the membrane. At the reported temperature, the device power consumption was ~100mW. The conductance variation of the sensing layer was measured by biasing the SnO_2 film at V_{Bias}= 1V. The relative conductance variation, $\Delta I/I_0$, of the sensing layer was measured.

Tetra t-Butyl Zinc Phthalocyanine was synthesized and purified in INQUIMAE, following the traditional procedures for metal phthalocyanines. The films were deposited either on a massive Si substrate covered with 1 micron of SiO_2 or upon a 200 nm thick Si_3N_4 membrane. A sub-group of the former 100 nm SnO_2 films undergo a second evaporation in vacuum of 30nm of t-Bu-ZnPc on the SnO_2 layer kept at 300 °C.

4. Modeled electronic nose

We designed a modeled nose with four micromachined sensors. The measured responses for each sensor and gas were modeled, through a fitting procedure, to enlarge the space of available data and analyze the possibility of gas identification. Noise was added to the modeled response by defining within a 20% of uncertainty the fitting parameters. The overall nose response reduced by the Principal Component Analysis technique to the first two components (96% of the total amplitude) is shown in figure 5.

Pure gas is identified through position in the p_1-p_2 plane. Gas concentrations are determined through a feed-forward neural network separately trained for each gas with back propagation procedure. The architecture has four input units, which are the normalized current detected by each sensor, three hidden units and one output unit that gives the concentration. Results for gas concentrations are plotted in figure 6 showing a relative good performance. We are analyzing other alternatives through variations in the number and type of sensors in the nose and optimization of the overall pattern recognition scheme.

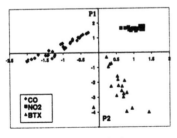

Figure 5: First two principal components   Figure 6: Assignment of concentration

5. Conclusions

We have shown that it is possible to determine the concentration of a given gas of environmental concern when detected in the presence of other sensible gases. This was based on the response of specific sensors (SnO_2) for which information from calibration data sheets were available. A procedure to construct a modeled response when two or more of the sensible gases are simultaneously present was proposed.

We developed micromachined sensors, very promising for atmospheric pollution applications, whose sensitivities lie at the few part per million-concentration range and we have enlarged the available experimental data, through modeling each sensor response, and analize the possibilities of an electronic nose based on these thin film sensors, to identify relevant gases of environmental concern and its concentrations

6. References

[1]   Dori L., Maccagnani P., Cardinali G.C., Fiorini M., Sayago I., Guerri S, Rizzoli R.and Sberveglieri G. 1997 Eurosensors XI , Warsaw, Poland,.

[2]   Gardner J.W. and Bartlett P. N. 1996 Sensors Actuators B 33 60-67.

[3]   Negri R.M., Reich S.L., Peltzer G., Romain A. C., Nicolás J., Lamagna A. and Dori 1999 L.. "Olfaction and electronic nose", Tubinguen, Germany, 372-375.

Section 3: Data Processing
Paper presented at the Seventh International Symposium on Olfaction and Electronic Noses, July 2000

165

Fuzzy Logic Processing in Combined Carbon Monoxide and Methane Domestic Gas Alarm

T. Šundić, A. Perera, S. Marco, A. Pardo, A. Ortega and J. Samitier

Instrumentation and Communication Laboratory
Departament d'Electrònica, Universitat de Barcelona
C/Martí i Franqués, 08028 Barcelona, Spain
Tel: +34 93 402 90 70, Fax: +34 93 402 11 48, E-mail: teodor@el.ub.es

Abstract

A complete description of combined gas alarm is presented. The alarm is based on two commercial metal-oxide gas sensors, a temperature sensor and Fuzzy processing engine implemented in simple 8-bit Toshiba micro controller. The algorithms for gas classification as well as for concentration estimation, both Fuzzy orientated, are detailed. Results obtained from exhaustive sensor and detector tests are presented.

1. Introduction

The demand of Natural Gas and Carbon Monoxide gas alarms for residential use has been increasing across Europe, North America and Japan. The main concerns are the explosions risks caused by gas leaks or Carbon Monoxide poisonings caused by incomplete combustion or the reflow of exhaust gases from gas burners in residential areas.

However, up to now low cost commercial products are usually based on metal oxide gas sensors. They present serious deficiencies in terms of selectivity and stability that result in a high percentage of false alarms. This causes client distrust in the alarm signal and fatal consequences can occur. Alternatively, lack of alarm signal in dangerous situations may also occur. Both cases represent a major failure in a security-oriented system. A multisensor approach and digital compensation algorithms running in an integrated microcontroller may correct some of these drawbacks.

In this context, it is the aim of this work to design and build a reliable and cost-effective detector using available sensor technologies and integrated soft-computing algorithms. This smart alarm should be commercially available shortly. In the following section, a brief description of the detector conception is given.

2. Detector structure and functioning aims

As mentioned in the introduction, the problem of nuisance alarms is one of the major problems for this type of gas detectors. The number of interferent gases that can influence the sensor signal as well as the temperature and humidity dependence are the most frequent reasons for their occurrence. Therefore, apart from the sensors with good sensitivity to CO and methane, we have considered adding one more sensor sensitive to majority of the gasses that can be found in domestic premises. A room temperature sensor also forms part of the detector and is used for the temperature correction of the sensor signals.

Fuzzy inference systems for gas classification and concentration estimation have been developed and implemented in 8-bit Toshiba microprocessor. The use of Fuzzy engines is based on their possibilities for more advanced processing needed in our case and also in the simplicity of implementing such a system in the selected microprocessor. One of the objectives proposed was the self-calibration of the detector based on very few calibration points. Considering the high cost of this process, the number of calibration points had to be very small while ensuring the good performance of the system. In our case, we have used five calibration points divided in the following way: three CO concentrations (50 ppm, 100 ppm and 300 ppm), one methane concentration of 5000 ppm and one ethanol concentration of 500 ppm. Instead of ethanol we could have also used some other interferent gas that the sensor is sensitive to, considering that the purpose is to eliminate the intevariability between devices in the absolute resistance.

3. Sensor selection and Test

As it is well known, one of the most important parts of any gas measuring system, if not the most important one, is the sensor. Therefore, the sensor choice is a basic part of the detector development. An initial program of exhaustive test of different pre-selected commercial sensors was planned to choose the ones with best characteristics. As our application combines detection of carbon monoxide and methane, we have considered sensors able to detect both gases. Lately, a number of manufacturers offer sensors that permit the simultaneous detection of CO and CH_4 (FIS, Microsens, MICS...). These sensors operate in a pulse mode changing the working temperature with a period of time of 10 to 15 s. CO is detected on low working temperatures ranging from 80°C to 200°C, while methane detection is usually performed on temperatures between 400°C and 500°C, although detection at the low temperature has been also proposed. . The presence of active filters enhance their selectivity characteristics, but on other side can bring problem of filter poisoning after a long presence of some gases like i.e. ethanol. Apart from the sensitivity and selectivity, characteristics like temperature and humidity dependence as well as inter-variability between sensors have also been considered as the important point in final choice.

In figure 1a and 1b we present the sensitivity and selectivity characteristics of the sensors tested in our laboratory, respectively. The FIS SB95 –11 model was the chosen model due to its characteristics. As can be seen in figure 1a, for both gases, CO and methane, the sensitivity is higher than for the rest of the sensors tested. On other hand, as shown in figure 1b, resistance clusters for CO and methane are well separated that is not the case for the other model presented in the figure. This can facilitate the classification of the gas present in the air and reduce the problem of false detection, i.e. false CO alarms in the presence of methane and vice versa.

FIS SP32 model was chosen for interferent detection.

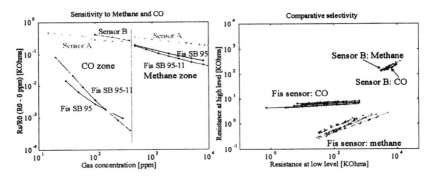

Figure 1: a) Sensitivity test of four different commercial sensors to methane and carbon monoxide b) Selectivity test of two commercial sensors to methane and CO

4. Fuzzy signal processing

The complete signal processing of the detector is based on different fuzzy routines and is performed in three phases. At first place, the temperature dependence of the sensor is corrected using the temperature sensor. The correction curve has been calculated after the exhaustive temperature test carried out in our laboratory, covering the whole range of interest from –10°C to 40°C. For the CO, the reference measurement is considered at the concentration of 60 ppm, while in the case of methane the concentration was 3000 ppm. The correction parameters for different temperatures are stored in look up table. Therefore, the temperature influence is reduced for all concentrations, but completely corrected only for the above ones.

4.1 Data

For the purpose of the classifier training, which is detector independent and consequently it does not depend on the detector calibration, a total of 112 data samples from four detectors has been used. Each detector combines a set of 28 samples divided in the following way: 4 CO concentrations (30, 50, 100 and 300 ppm), 4 methane concentrations (1000, 2000, 5000 and 8000 ppm), 14 air samples, 4 ethanol concentrations (500, 1000, 1500 and 2000ppm) and one concentration of methanol (1000 ppm) and acetone (200ppm). Air is represented in different humidity and temperature conditions, from 30% to 95% of relative humidity and from –10°C to 40°C, covering the whole range proposed by [2]. Four concentrations of ethanol have been added because it is considered one of the most important influences to the proper detector functioning. At the beginning, some other interferent gases like (SO2, 2-propanol, ammoniac, toluene etc.) have also been treated but due to the low sensor response they were discarded from the last training data set. All algorithms were validated by two other detectors with a total of 56 samples.

4.2 Classifier design

A fuzzy classification algorithm that can distinguish between four classes is developed. Apart from the CO and methane classes, the two other classes are air and interferent gases. The system has three inputs: SB95-11 resistance at high temperature (methane sensitivity), SB95-11 resistance at low temperature interval (CO sensitivity) and SP 32 response.

A sugeno type fuzzy inference system based on the subtractive clustering algorithm was developed. This method assumes each data point as a potential cluster centre and based on the density of surrounding data points calculates the possibility that each data would define the cluster centre [3]. Once the centres are selected, correspondent sets of fuzzy rules and membership functions are extracted and optimised. During the training process the radius, that specifies the cluster centres range of influence in each of the data dimensions, is optimised being the percentage of well-classified samples the objective function. To each cluster centre one fuzzy rule is associated and in that way dividing the feature space into the number of partitions equal to the number of centres. In our case, the number of centres was four, coinciding with the number of classes. Still, it is not the general rule, and the number of centres can vary depending on the radius of the inputs and, of course, on data. In figure 2 we present the position of the fuzzy centres in logarithmic data space in all three projections together with data. In the same figure, the curves around the centres are representing the maximum of the four membership functions belonging to each centre giving us idea of the fuzzy partition of the input space.

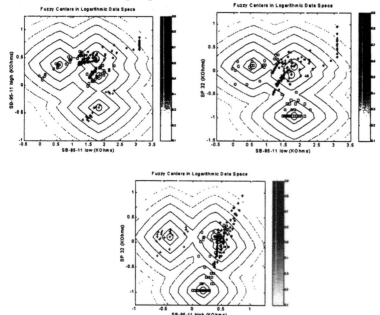

Figure 2. Fuzzy centres (O) in logarithmic data space. Data is represented by: CO (o), Methane (+), Air (*) and interferent gases( ).

The input membership functions were triangular type due to the constraint of the microprocessor fuzzy engine that permits only triangular and trapezoidal input functions. The defuzzification method is weight average and the output membership functions are singleton type.

The error in classification process, both in train and test sets, is 8%. The error appears because some interferent gases are classified as air due to a low sensor response to these gases. Nevertheless, this error does not have any relevance in the correct functioning of the gas alarm and can be omitted.

4.3 Concentration estimation

Once the gas is classified, the concentration estimation is performed. Another fuzzy system has been designed to approximate the sensor response curve. For all three gases (CO, methane and ethanol) five reference positions (corresponding to five concentrations of the gas) of the input fuzzy sets are fixed based on the previous test performed in the laboratory. For CO the concentrations are 20, 50, 100, 300 and 400 ppm, for methane they correspond to 200, 500, 5000, 8000 and 10000 ppm while in the case of ethanol the references are fixed to 200, 500, 1000, 1500 and 2000 ppm. A set of triangular membership functions, defined from the calibration points, covers the input universe of discourse. In CO case, from three calibration points (50, 100 and 300 ppm), and the reference values, other two (20 and 400 ppm) are extrapolated. In methane case from just one calibration point at 5000 ppm other four are extrapolated, similar as for ethanol where from 500 ppm calibration point the rest is extrapolated. The extrapolation is based on the hypothesis of the potential dependence between the resistance change and the concentration.

Resistance values are fuzzified and processed by a fuzzy engine that is based on simple *if-then* fuzzy rules. Finally, defuzzification based on weight average to a gas concentration space is performed. A fixed set of triangular output membership functions is defined whose width is optimised to fit the curvature of the sensor response. In figure 3a we present the fuzzy approximation of the calibration curve compared to linear interpolation. It can be observed that with only three calibration points a very low conformity error is found.

In figure 3b we show the response to the CO concentration. The maximum relative error is less then 10%, which can be considered as acceptable. The response steps that can be noticed at high concentrations are due to the quantification steps of the 8 bit A/D in the Toshiba microprocessor. At low concentrations, this steps are small and are growing as the concentration gets higher. In methane case the error is up to 20% relative error depending on the ambient conditions. This is due to just one calibration point at 5000ppm that makes difficult precise response curve adjustment.

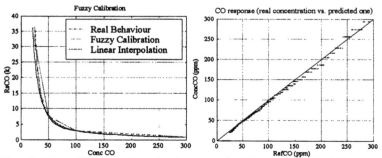

Figure 3. a) Fuzzy CO concentration estimation compared with linear interpolation, b) Detector response to CO concentration estimation

However, the methane case is not critical as the CO one. According to international standards (ie. [2], etc.), permitted error is ± 1250ppm. Fixing the alarm point at 5000 ppm of methane, the 20% relative error (1000 ppm) does not overpass the permitted levels.

Apart from the results presented above, exhaustive test over temperature and humidity dependence has been carried out at critic gas concentrations and the results obtained are satisfactory. The maximum relative error that raises from the humidity dependence at the 50 ppm of CO, the concentration in which the most signal fluctuation was observed, was less 8%.

At figure 4 we present the final outlook of the detector.

Figure 4. Final outlook of the detector

5. Conclusions

In this paper, we have showed that is possible to obtain reliable gas detection device with low percentage error in concentration estimation. We have also proposed a way to minimise the false alarms problems. We hope that in the near future, after performing some new tests, we will be able to treat the problem of temporal drift that we consider one of the last steps in the development of this device.

Acknowledgements:

This project has been partially funded by Spanish CICYT Project TIC 98-0987-C03-03. Funding from Gas Natural SDG SA. and MCC is also acknowledged.

References:

[1] T. Sundic, S. Marco, A. Pardo, M. Ortiz and J. Samitier, "Quantitative Signal Processing Algorithms for Low Cost Methane and Carbon Monoxide Detectors", Proceedings of the IMTC-99 Venice, Vol.1, pp. 424-428, May 1999.
[2] P.K Clifford, M.G. Dorman, GRI, "Test Protocols for Residential Carbon Monoxide Alarms, Phase 1", September 1996.
[3] S. Chiu, "Fuzzy Model Identification Based on Cluster Estimation", Journal of Intelligent & Fuzzy Systems, Vol.2, No. 3, Sept. 1994.

Medical/Microbial

Section 4: Medical/Microbial
Paper presented at the Seventh International Symposium on Olfaction and Electronic Noses, July 2000

173

Effectiveness of an Electronic Nose for Monitoring Bacterial and Fungal Growth

S S Schiffman,[1] D W Wyrick,[1] R Gutierrez-Osuna,[2] H T Nagle[3]

[1]Duke University Medical School, Box 3259, Durham, NC 27705 USA;
[2]Wright State University, 401 Russ Engineering Center, Dayton, OH 45435;
[3]North Carolina State University, Box 7911, Raleigh, NC 27695

Abstract. Growth of microbial organisms such as bacteria and fungi generates volatile organic compounds and fixed gases. An electronic nose consisting of 15 metal-oxide sensors (NC State E-Nose) was used to detect and classify bacteria and fungi. Three preliminary experiments were conducted with the electronic nose using odorous stimuli related to microbial contamination. The results suggested that the NC State E-Nose could classify bacteria, fungi, and associated volatile organic compounds. A further experiment was performed to detect and classify five fungi commonly found in indoor environments. These fungi were *Aspergillus flavus, Aspergillus niger, Penicillium chrysogenum, Cladosporium cladosporioides,* and *Stachybotrys chartarum.* The fungi were cultured on two types of media, Potato Dextrose Agar (PDA) and Czapek-Dox Agar. The NC State E-nose was capable of discriminating among these fungi with up to 96% accuracy.

1. Introduction

Growth of bacteria and fungi on organic matter generates a broad range of volatile organic compounds and fixed gases. Wessén and Schoeps [1] and Sunesson et al. [2] showed that the presence of certain volatile organic compounds can be used as an indicator of the presence and identity of microorganisms. Holmberg [3], in a dissertation at Linköping University in Sweden, used an electronic nose with 15 sensors to classify 5 types of bacteria (*Escherichia coli, Enterococci* sp., *Proteus mirabilis, Pseudomonas aeruginosa,* and *Staphylococcus saprophyticus*). The 15 sensors included 9 MOSFETs, 4 Taguchi type sensors, 1 carbon dioxide sensor, and 1 oxygen monitor. The volatile compounds generated by the bacteria were sampled from agar plates. The results suggested that this E-nose could successfully classify *Escherichia coli* and *Enterococci* sp. but was less successful with the other bacteria.

Gardner et al. [4] used an electronic nose that contained six commercial metal oxide sensors, a temperature sensor, and a humidity sensor to predict the class and growth phase of two types of bacteria, *Escherichia coli and Staphylococcus aureus*. The six sensors were designed to detect hydrocarbons, alcohols, aldehydes/heteroatoms, polar molecules, and nonpolar compounds. The best mathematic model identified 100% of the unknown *S. aureus* samples and 92% of the unknown *E. coli* samples.

Other studies have also found that bacteria can be discriminated using an electronic nose. In an evalutation of 7 bacterial strains, Vernat-Rossi et al. [5] were able to correctly discriminate 98% of a training set with a cross-validation estimate (test set) of 86% using 6 semiconductor gas sensors. Parry et al. [6] found that swabs from chronic venous leg ulcers with haemolytic streptococci could be discriminated from those without bacteria using an electronic nose with polymer sensors. Studies at AromaScan PLC [unpublished data from Dr. Krishna Persaud] showed that polymer sensors performed well in discriminating multiple samples of 5 different types of bacteria.

Keshri et al. [7] used an electronic nose consisting of 14 polymer sensors to classify six spoilage fungi (four *Eurotium* sp., a *Penicillium* sp. and a *Wallemia* sp.). The headspace was sampled after 24, 48, and 72 hours of growth. The electronic nose discriminated the fungi at the 24-hour mark (prior to visible signs) with an accuracy of 93%. The best results occurred at the 72-hour mark.

2. Objectives of the present study

The purpose of the present study was to extend our current understanding of the capacity of an electronic nose to indentify and classify microorganisms including bacteria and fungi. When conditions are favorable and a nutrition source is present, microbial organisms such as fungi and bacteria can grow almost anywhere. Microorganisms have been shown to generate volatile organic (VOCs) while metabolizing nutrients, and these VOCs have been used as indicators of microbial growth. Microorganisms not only generate airborne contamination in the form of VOCs, but also generate toxins and propagules including conidia (spores) and bacterial cells. When microoganisms infest buildings, they can create a potentially hazardous environment. Individuals exposed to environments containing high concentrations of airborne contaminants from microbial organisms report symptoms including eye and sinus irritation, headaches, nausea, fatigue, congestion, sore throat, and even toxic poisoning. Sick Building Syndrome (SBS), which includes health symptoms arising from poor indoor air quality, have been correlated with the presence of fungi [8]. A study of two houssholds reporting indoor environmental complaints correlated the presence of excessive VOC's with the presence of fungal contamination [9]. Typical, signs of microbial contamination include water damage, high levels of humidity, and physical presence. However, these signs are not always present and therefore cannot be used as sole indicators of microbial contamination.

Current methods for detecting microbial contamination include visual inspection, air and material sampling with culture analysis, and air sampling coupled with gas chromatography/mass spectrometry [10,11]. These methods, however, can be inconclusive, time consuming, and expensive. There is a need for rapid detection of the presence of microbial contamination in order to minimize its impact. The studies described in this paper indicate that the use of an electronic nose can reduce the time required to detect, discriminate, and identify microbial contamination.

3. NC State electronic nose

An electronic nose instrument was designed and constructed at North Carolina State University [12,13,14] that uses an array of metal oxide sensors for measuring odor in air samples. It consists of a sampling unit, a sensor array, and a signal processing system. The sampling unit, which consists of a pump and a mass flow controller, directs the air sample containing the odorant under investigation across the sensor array. The current configuration allows for sampling from a set of 12 odorants, a reference sample (filtered odorless dry ambient air), and a washing agent (ambient air bubbled through a 2% butanol solution). The sensor array is comprised of 15 different metal oxide sensors each producing a different response pattern for the odorant under investigation. Twelve of the 15 metal oxide sensors are manufactured by Capteur (Didcot, UK) and include sensors for isopropyl alcohol, toluene, hydrogen sulfide, nitrogen dioxide, chlorine, butane, propane, hydrogen, carbon monoxide, heptane, ozone and general VOCs. The remaining three metal oxide sensors are produced by Figaro USA (Glenview, IL) and include a methane, a combustible gas, and a general air contaminant sensor. The overall combination of the response patterns defines the odorant under investigation. All of the sensor response patterns are digitized and recorded using a National Instruments® Data Acquisition Card controlled by LabVIEW®. The data are analyzed with MATLAB® using signal processing algorithms developed by Kermani [12] and Gutierrez-Osuna [13].

The raw data are first compressed using windowing functions that produced a set of four features for each sensor. Linear discriminant analysis (LDA) is then applied to the compressed data to maximize class separability. Sixty percent of the compressed data is randomly selected to form a training set for the classification algorithms. K nearest neighbors (KNN) and least squares analysis (LS) are both employed to classify the remaining 40% of the compressed data [15]. This method is performed a hundred times and the average score is used for the final classification score.

4. Preliminary experiments with NC State electronic nose

Three preliminary studies of the effectiveness of the NC State E-Nose were conducted on odorous stimuli related to microbial contamination. These studies included: 1) discrimination among ten bacterial organisms, 2) discrimination among five volatile organic compounds known to be generated during fungal and bacterial growth, and 3) discrimination among three fungal organisms. These studies described below showed that the NC State E-Nose is effective in discriminating odors associated with microbial contaminants. This prompted a final experiment in which the three fungal organisms from the second preliminary experiment along with two other fungi and two controls were investigated.

4.1. Preliminary bacterial experiments

Ten bacterial standards were grown on agar using the Prompt^{TM} inoculation system (Becton Dickinson), and the headspace was sampled from the petri dishes 24 hours after inoculation. The ten bacteria included: *Escherichia coli, Enterobacter cloacae, Group D enterococcus, Citrobacter freundii, Klebsiella pneumoniae, Staphylococcus agglutinin negative, Staphylococcus aureus, Serratia fonticola, Proteus mirabilis, and Pseudomonas aeruginosa.* The stimuli were sampled for 60 seconds with a 30 second wash cycle followed by a 240 second reference sample. The best classification algorithm yielded a correct percent classification of 82.5% (using leave-one-out cross-validation). The confusion matrix for these bacteria is given in Table 1. The statitical technique of linear discriminant analysis

(LDA) was applied to the compressed data to maximize class separability followed by a K Nearest Neighbor classifier. The scatter plot generated by LDA is given in Figure 1. The LDA scatter plot suggests very good separation between classes, but the 82.5% validation rate indicates that the classifier overfits the training data, since only 4 sniffs per culture were collected. This prompted us to collect a larger data set for the final experiment reported in section 5 of this article.

Table 1: Confusion matrix from bacteria with percentages of classification

Predicted Class	True Class									
	1	2	3	4	5	6	7	8	9	10
1 *Escherichia coli*	75	0	0	0	0	0	0	0	0	0
2 *Enterobacter cloacae*	0	100	0	0	0	0	0	0	0	0
3 *Group D enterococcus*	25	0	50	0	0	50	0	0	0	0
4 *Citrobacter freundii*	0	0	0	75	25	0	0	0	0	0
5 *Klebsiella pneumoniae*	0	0	0	25	75	0	0	0	0	0
6 *Staphylococcus agglutinin negative*	0	0	50	0	0	50	0	0	0	0
7 *Staphylococcus aureus*	0	0	0	0	0	0	100	0	0	0
8 *Serratia fonticola*	0	0	0	0	0	0	0	100	0	0
9 *Proteus mirabilis*	0	0	0	0	0	0	0	0	100	0
10 *Pseudomonas aeruginosa*	0	0	0	0	0	0	0	0	0	100

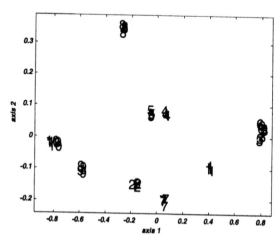

Figure 1: LDA scatter plot of bacterial data

4.2 *Preliminary volatile organic compound experiments*

In the second preliminary experiment the sensitivity of the NC State E-Nose to three VOCs generated by fungi was evaluated. Three VOCs selected for the experiment were ethanol, 3-pentanone, and 2-methyl-1-propanol. Serial dilutions of each chemical were prepared with dionized water to obtain eight concentrations. The headspace of each concentration was randomly sampled 3 times daily for 5 days from 30 ml amber bottles containing a 15 ml of solution. The headspace was drawn through a small hole in the lid of the bottle using a PVC tube. The results from the preliminary experiments on the volatile organic compounds indicated that detection thresholds for the NC State E-Nose were between 0.003% and 0.002% for the sample dilutions (see Table 2). The percent correct classification for all the chemical samples (and the diluent control) into their respective chemical classes was obtained by training a separate classifier at each concentration (see Table 3). These data suggest that the NC State E-Nose is capable of detecting and discriminating volatile compounds known to be generated during fungal growth.

4.3 *Preliminary fungal experiments*

Three fungi, *Aspergillus flavus*, *Aspergillus niger*, and *Penicillium chrysogenum*, were incubated at 28°C on 100 mm petri dishes containing Potato Dextrous Agar (PDA). The headspace above each fungus was sampled through a small hole in the center of the lid of the petri dish and an inline one micron filter for removing conidia (spores). The percent correct classification for the three fungi continued to improve through the seventh day of growth (see Table 4). It can be seen that 90% classification was obtained using KNN on days five and seven combined. The conclusions from this and the other two preliminary experiments indicated that the NC State E-Nose is capable of discriminating odors generated by microbial samples.

Table 2: Percent correct detection of the headspace samples from the serial dilutions

Chemical Type	Classification Method	Percent Concentration of Serial Dilution				
		0.050%	0.013%	0.006%	0.003%	0.002%
Ethanol	KNN	99%	95%	97%	88%	74%
	LS	99%	95%	97%	88%	74%
2-Methyl-1-Propanol	KNN	99%	92%	84%	81%	72%
	LS	99%	92%	83%	82%	72%
3-Pentanone	KNN	100%	99%	99%	98%	85%
	LSS	100%	99%	99%	98%	85%

Table 3: Percent correct discrimination of headspace samples of 3 VOCs and a diluent control for each serial dilution

Classification Method	Percent Concentration of Serial Dilution				
	0.050%	0.013%	0.006%	0.003%	0.002%
KNN	97%	94%	84%	84%	76%
LS	92%	83%	73%	75%	72%

Table 4: Percent correct classification for 3 fungal species grown on PDA

Classification Method	Day of Growth					
	0	1	3	5	7	5&7
KNN	38%	39%	69%	75%	79%	90%
LS	39%	39%	56%	69%	79%	82%

5. Final experiment with NC State electronic nose

An additional experiment was performed on a larger data set of fungi with improved instrumentation and sampling techniques. Five fungi (*Aspergillus flavus, Aspergillus niger, Penicillium chrysogenum, Cladosporium cladosporioides, and Stachybotrys chartarum*) were incubated at 28°C on 150 mm petri dishes containing Potato Dextrous Agar (PDA), a complex media rich in nutrients, and Czapek-Dox Agar (CZ), a minimal media. These two types of media were used in order to provide two different growth environments and to produce different growth rates (see Figure 2). Twenty four petri dishes of each media were inoculated with 0.5 ml of an individual spore suspension containing 10,000 condia/ml from each fungus respectively. The suspensions were prepared using a Spencer hemacytometer with improved Neubauer ruling. An autosampler was constructed to sample uniform air volumes at controlled intervals. Using the autosampler, air samples from the headspace of each petri dish containing one species on each media were randomly sampled ten times each after 24 hours and every other day thereafter for two weeks. The headspace above each fungus was sampled through a small hole in the center of the lid of the petri dish using a PVC tube and an inline two micron filter for removing conidia (spores).

The data were analyzed using two classification protocols. In the first protocol, the data were grouped into 12 classes: 5 fungal species grown on PDA and CZ respectively plus 2 controls (the two media PDA and CZ without fungal growth). The results are shown in Table 5. After 24 hours of growth, the percent classification was 90% for K nearest neighbors, and 76% for least squares. Classification for 12 classes reached a maximum after 5 days of growth, with an accuracy of 96% for K nearest neighbors and 94% for least squares. After day 5, the percent classification began to decrease slowly. By day 15, the percent classification was reduced to 89% for K nearest neighbors and 69% for least squares.

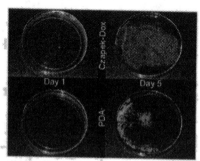

Figure 2 *Stachybotrys chartarum* growing on PDA and Czapek-Dox at day 1 and 5

Table 5: Percent classification for 12 classes (5 fungal species on two different media and 2 control media)

Classification Method	Day of Growth							
	1	3	5	7	9	11	13	15
KNN	90%	91%	96%	94%	89%	93%	93%	89%
LS	76%	90%	94%	90%	93%	86%	80%	69%

Table 6: Percent classification of 7 classes (5 fungal species and two control media)

Classification Method	Day of Growth							
	1	3	5	7	9	11	13	15
KNN	89%	90%	94%	93%	89%	94%	94%	92%
LS	79%	88%	93%	91%	95%	90%	92%	86%

In the second classification protocol, the data were grouped into 7 classes: 5 fungal species (independent of media used for growth) plus 2 controls (the two media PDA and CZ without fungal growth). In other words, each of the fungi grown in PDA and CZ were combined into a single class. After 24 hours of growth, the percent classification was 89% for K nearest neighbors, and 79% for least squares. Classification reached a maximum after 5 days of growth, with an accuracy of 94% for K nearest neighbors and 93% for least squares. After day 5, the percent classification oscillated around an average percent classification of 92% with a standard deviation of 2%. The results are shown in Table 6. This final experiment using 5 fungi resulted in improved classification beyond that obtained for 3 fungi in the preliminary results.

6. Conclusions

Both the preliminary studies as well as the final experiment with 5 fungi showed that the NC State E-nose can detect and classify microorganisms on the basis of volatile emissions. The improved classification in the final experiment with fungi was independent of the media used to grow the fungi. Furthermore, correct classification was achieved with the NC State E-Nose at 24 hours of growth. Thus the NC State E-Nose has potential to be used for early detection of microbial contamination.

7. Reference

[1]   Wessén B and Schoeps K-O 1996 Analyst 121 1203-1205
[2]   Sunesson A-L et al. 1995 Appl. Environ. Microbiol. 61 2911-2918
[3]   Holmberg M 1997 Depart. Phys. Meas. Tech. Linköping University Sweden
[4]   Gardner J W et al. 1998 Meas. Sci. Tech. 9 120-127
[5]   Vernat-Rossi V et al. 1996 Sensors and Actuators B 37 43-48
[6]   Parry A D et al. 1995 J. Wound Care 4 404-406
[7]   Keshri G et al. 1998 Lett. Appl. Microbiol. 27 261-264
[8]   Ahearn D G et al. 1996 J. Indust. Microbiol. 16 280-285

[9]   Ström G et al. 1994 Health Implications of Fungi in Indoor Environments (Amsterdam: Elsevier) 291-305

[10]   Schiffman S S et al. 2000 Agr. Forest Meteorol. in press.

[11]   Pasanen A L et al. 1992 Int. Biodeter. Biodegrad. 30 273-283

[12]   Kermani B G 1996 Depart. Electrical Computer Engineering North Carolina State University USA

[13]   Gutierrez-Osuna R 1998 Depart. Electrical Computer Engineering North Carolina State University USA

[14]   Wyrick D W 2000 Depart. Electrical Computer Engineering North Carolina State University USA

[15]   Duda R O and Hart P E 1973 Pattern classification and Scene Analysis (New York: Wiley)

Section 4: Medical/Microbial
Paper presented at the Seventh International Symposium on Olfaction and Electronic Noses, July 2000

181

Investigation of the Growth Characteristics of *E. Coli* using Headspace Analysis

R Esteves de Matos, D J Mason, C S Dow and J W Gardner*

Department of Biological Sciences, and School of Engineering*, University of Warwick, Coventry CV4 7AL, UK.

Abstract. The Agilent 4440 chemical sensor has been used to study several aspects of the growth of *Escherichia coli*. This instrument consists of an automated headspace sampler coupled to a quadrupole mass spectrometer and is designed to analyse the headspace of samples, in this case bacterial cultures, for their volatile composition over the mass range 46 to 550 daltons. The volatile signature has been investigated with respect to each growth phase as follows; **lag**: initiation of growth and macromolecular synthesis, **exponential** (or log): differentiation of signature with respect to metabolism (low mass ranges) and bacterial speciation (high mass ranges), **stationary**: categorisation of the bacterial population into growing, stressed and dormant cells. These objectives have been addressed using electrical flow impedance to determine cell size and number, and fluorescence to determine the physiological state of individual cells within the bacterial population. Principle component analysis shows that the volatile signature can be used to discriminate between the different growth phases with most of the variance contained within the mass range of 46 to 100 daltons.

1. Introduction

Discrimination between viable, dormant and dead micro-organisms still remains a fundamental problem of microbiology even with advanced molecular techniques, such as nucleic acid probing. Such techniques are constrained, and are at present only capable of nucleic acid detection regardless of the physiological state of cells. Mono-cultures of bacteria in batch and continuous culture are much more heterogeneous than was previously assumed both genetically and physiologically, and as a result such cultures consist of several sub-populations differing in viability, activity and cellular integrity. *In situ* analysis of the physiological activity of individual cells which comprise a bacterial population, with quantitative and qualitative evaluation of their physiological states, i.e. reproductive, stressed, dormant and dead, is essential for our understanding of basic microbiological processes and for the development and validation of non invasive analyses (e.g. e-nose technology). A further problem is the temporal expression of particular genes e.g. verocytotoxin genes of *E.*

coli O157:H7. Recent evidence suggests that enhanced expression of the verocytotoxin genes is observed when cells are stressed/dormant compared to the levels expressed in active/reproductive cells. A situation that is of obvious clinical significance. It is important, therefore, that we have a clear understanding not only of microbial growth, but also of the expression of specific genes in the physiologically distinct sub-populations within a microbial culture.

2. Materials and Methods

2.1 Bacterial strains and culture conditions

Escherichia coli NCTC 10538 (a K-12 strain) was the bacterial strain used throughout this study. Cultures of *E.coli* were incubated aerobically in Luria Bertani broth (LB) at 37°C shaking at 200 rpm or, for starvation conditions, in minimal media M9 (Oxoid, UK) containing 0.05% (w/v) glucose as a limiting carbon source.

Appropriate volumes of an overnight culture of *E. coli* in LB, incubated at 37°C, were inoculated into fresh LB to give a final concentration of 10^6 cells ml^{-1}. The flasks were incubated at 37°C with shaking and samples were taken at regular intervals from which aliquots were removed for Agilent 4440 analysis (see later), flow cytometric analysis using SYTO 13 (Molecular Probes, Inc) and propidium iodide (flow cytometry) as the fluorescent stains, plate counts, particle sizing and total particle counts (*CellFacts* I, Microbial Systems Ltd).

For starvation experiments an overnight culture of *E. coli* (stationary phase) grown in M9 containing 0.05% (w/v) glucose, which had been incubated at 37°C, was inoculated into fresh M9 medium containing 0.05% (w/v) glucose and incubated at 37°C with shaking. Sampling was carried out as described above.

2.2 Flow cytometry

SYTO 13, which was supplied as a 5mM solution in dimethyl sulfoxide, was diluted in sterile distilled water to give a working solution of 100 μM. This is a permeable stain, i.e. it will stain all cells green. Propidium iodide was dissolved in sterile distilled water to give a stock solution of 100μg ml^{-1}. This dye will only enter dead cells, staining them red. Both are nucleic acid dyes.

An aliquot from each sample was incubated for 10 minutes at room temperature with both SYTO 13 and propidium iodide to a final concentration of 5 μM and 10 μg/ml, respectively, before analysis in a flow cytometer.

All flow cytometric analyses were carried out using a Bryte HS flow cytometer (BioRad, Hemel Hempstead, UK) fitted with a mercury-xenon arc lamp. The instrument is equipped with two light scatter detectors (<15° and >15°, measuring forward and side scatter, respectively) and three fluorescence detectors detecting appropriately filtered light at green (515 to 565 nm; to detect SYTO 13 fluorescence), orange (565 to 605 nm) and red (>605 nm; to detect propidium iodide fluorescence) wavelengths. Dye excitation was achieved at 470 to 490nm. The beam splitter was designed to allow detection of light emission at 520nm and above. At least 5,000 events were recorded for each sample and the experiments were conducted in triplicate. Instrument performance was monitored and adjusted as appropriate using fluorescent 1.5 μm diameter calibration beads (BioRad) and the instruments own electronic compensation was used to adjust any overlap of fluorescence between green and red fluorescence parameters. Positive controls were also used for propidium iodide fluorescence (heat killed cells) and for SYTO 13 fluorescence (live cells).

2.3 Head space analysis

Agilent 4440 chemical sensor (Agilent Technologies Inc., USA) comprises a headspace autosampler and quadrupole mass spectrometer, see Figure 1. It allows for sample classification and screening of multiple applications on the same system by analyzing the volatile constituents in the headspace of samples held in sealed vials.

Here samples were taken at regular intervals during growth of *E.coli* in LB and 1.5 ml was injected into sterile sealed 10 ml vials. The vials were crimped manually and then placed in the Agilent 4440 chemical sensor set up with the parameters in table 1.

Figure 1. The Agilent 4440 chemical sensor used to analyse the headspace of bacteria.

Table 1 Agilent 4440 Chemical Sensor: experimental settings.

Parameter:	Value set
Oven temperature	80°C
Loop temperature	90°C
Transfer line	100°C
HS cycle time	4 min
Vial Equilibration Time	12 min
Loop fill time	0.15 min
Loop Equilibration Time	0.02 min
Pressurise time	0.30 min
Inject time	0.30 min
Carr. Press. PSI	4.5 PSI
Vial Press PSI	14 PSI
Temperature controller	110°C
Ionisation Gauge	6.6×10^{-5}
Solvent delay	0.45
Running time	0.75

2.4 Plate counts (Colony forming units)

Plate counts were carried out by diluting samples of *E.coli* in phosphate buffered saline (PBS) and spreading appropriate dilutions on nutrient agar plates (Oxoid, UK). Colonies were counted after incubating the plates at 37°C for 24 hours.

2.5 Total particle counts and cell size measurements

Total particle counts and individual cell sizes (particle distribution profiles) were obtained by using a *CellFacts* I instrument (Microbial Systems Ltd) which is an electronic particle counter. This instrument reports the volume of individual cells/particles in an electrolytic solution by measuring the change in impedance across a 30 μm × 80 μm orifice as individual particles transit the orifice.

3. Results

3.1 Determination of growth phase

Cell size is a reflection of the growth state of the microbe. In the lag phase there is a significant increase in cell size without cell division (i.e. there is macromolecular synthesis). During the log phase, the growth rate correlates with cell size that remains relatively constant. With the onset of stationary phase (nutrient limitation) a decrease in cell size is observed with all cells ultimately achieving the same cellular dimensions. It is therefore possible to relate easily the volatile signature to each of the specific growth phases via the cell size data. Figure 2 shows both the cell count and mean cell diameter during the first 8 hours of growth. This plot is used to define the growth phase and in

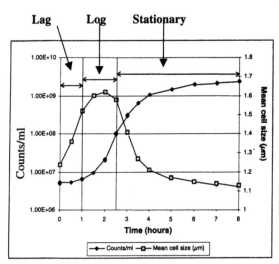

Figure 2. Changes in total particle count and cell size during growth of *E. coli* in Luria Bertani broth (LB) at 37°C as determined by electrical flow impedance (*CellFacts I, Microbial Systems*).

this run the lag/log boundary occurs at about 1 h and the log/stationary boundary at about 2.5 h. The volatile signature of each sample was then classified according to this rating.

3.2 Chemometric analysis of volatile mass signatures

The Agilent 4440 chemical sensor was used to analyse the volatile headspace of the *E.coli* samples over the full 20 day period. Figure 3 shows the mass spectrograph of samples taken within the first 8 hours and shows just two of the three growth phases with a typical lag phase signature plotted upwards and a typical stationary phase signature downwards for the sake of clarity. The mass range was set from 46 to 550 daltons. This figure shows that the typical mass ion abundances are modest with a total abundance of up to about 10^6 and so the matrix is relatively simple and the output should be reasonable linear. The existence of a significant peak does not mean that it contributes to the discrimination of growth phase, but the relative ratios of mass ions for each growth phase are important. However, it is hard to know their relative merit with 505 mass variables.

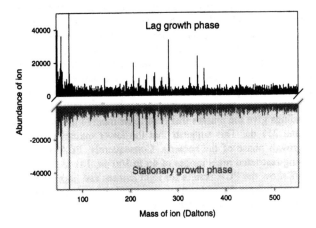

Figure 3. Typical mass spectrograms of the headspace of *E.coli* samples taken in the lag and stationary phases (note that the stationary growth phase data is inverted for the sake of clarity).

In order to classify the volatile signatures, we have applied principal component analysis (PCA) to the mass data using the commercial software Pirouette® version 2.7.

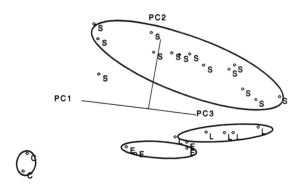

Figure 4. PCA analysis of the volatile headspace of *E.coli* samples taken by the Agilent 4440 unit. Masses: 46 to 550 da. Lag (L), Log (E), Stationary (S).

Figure 4 shows the first two principles components of all of the samples taken within the first eight hours and so covers the three growth phases as demonstrated by Figure 1. The other stationary samples appear to the left of the stationary cluster and have been removed for the sake of clarity. Two control samples (denoted by label C) can be seen as quite distinct.

From the PCA analysis, it is apparent that there is a relationship between the mass composition of the volatile headspace of *E.coli* and its different growth phases. Thus three growth phase clusters can be seen. There is also a clear movement to the left as the bacteria in the stationary phase ages. The extreme left sample is the oldest one at 8 h. Obviously the boundary between the different phases is incremental rather than discontinuous and so there is some overlap occurring.

The PCA loadings give the masses that contribute most to the variance and the first 5 are 46, 73, 48, 56 and 251 da. This suggests that the lower masses are most important in discriminating the growth phase of the bacteria. Consequently, PCA analyses were carried out using the following restricted mass ranges of 46 to 100 da, 101 to 200 da and 201 to 550 da. Figures 5, 6 and 7 show the PCA results for light masses and high masses. These plots it confirm the loadings result from the entire mass range and show that the mass range of 46 to 100 da can be used to discriminate the growth phase of *E.coli*.

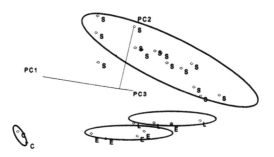

Figure 5. PCA analysis of the volatile headspace of *E.coli* samples taken by the Agilent 4440 unit. Masses: 46 to 100 da.

The highest loadings (i.e. most important masses) were now found to be 46, 73, 48, 56, and 47 (instead of 251) da. Note that the degradation of clustering that occurs when the mass range has been increased to only those ions with masses of over 200 da. Only the two control samples remain distinct while overlap is greatest between the lag (L) and log (E) samples. Thus the majority of the variance from the growth phase is expressed in the first 50 or so masses, i.e. 46 to 100 daltons.

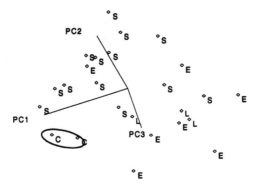

Figure 6. PCA analysis of the volatile headspace of *E.coli* samples taken by the Agilent 4440 unit. Masses: 101 to 200 da.

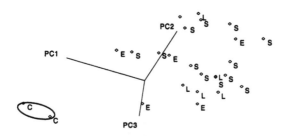

Figure 7. PCA analysis of the volatile headspace of *E.coli* samples taken by the Agilent 4440 unit. Masses: 200 to 550 da.

4. Conclusions and Further Work

Our results suggest that the volatile compounds produced in the headspace of *E.coli* grown in an aqueous nutrient solution can be used to determine its growth phase. Moreover, the low molecular weight compounds of 46 to 100 daltons were sufficient. We believe that the low molecular weight data can be correlated with cellular metabolism whilst the high molecular weight data is derived primarily from cellular components (e.g. membrane fatty acids), particularly at high oven temperatures i.e. >80°C.

It has been shown previously that it is possible to use the volatile signal to differentiate between bacterial species [1]. Having established the different growth phases, the investigation was broadened to include characterisation of the physiological states of individual cells comprising a population i.e. quantitative analysis of the component sub-populations. This was achieved using fluorescence probes. A green (permeable) and a red (impermeable) fluorescent stain were used for the characterisation of the physiological states of individual cells of *E. coli* (i.e. reproductive, stressed, dormant and dead). Live cells with intact bacterial membranes stain green, dead cells with damaged membranes stain red. The appearances of cells, which emit an orange fluorescence, are indicative of cells in a stressed state i.e. physiologically compromised.

A pattern was observed with respect to the changes in the fluorescence staining. Initially over 90% of the cells were within the population which stained green, and gradually during growth, these decreased whilst there was an increase in the numbers of orange and red staining cells which was dependent on the growth state and culture conditions, see Figure 8. From these data we are attempting to correlate the volatile patterns with the sub-populations in each growth phase (particularly during the stationary phase).

In addition to the analysis of sub-populations the expression of verocytotoxin in stressed cells is being investigated which follows on from previous work relating to the expression of hepatotoxins by cyanobacteria [2].

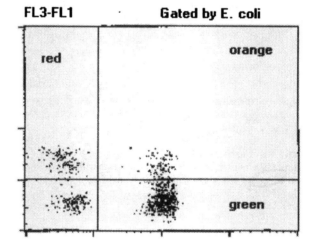

Figure 8.   Flow cytometry plot of forward scatter against fluorescence for *E.coli* challenged with two fluorescent probes.

Acknowledgements

The authors would like to thank Agilent Technologies (USA) for their support of this project re the Agilent 4440 chemical sensor; the Portuguese Government for sponsorship of Raquel Esteves de Matos and Microbial Systems Ltd for the use of *CellFacts*.

References

[1] J. W. Gardner, M. Craven, C. S. Dow and E.L. Hines, *Meas. Sci. Technol.*, 9 (1998) 120-127.
[2] H. W. Shin, E. Llobet, J. W. Gardner, E. L. Hines, C. S. Dow, *Proc. IEE, Science, Measurement & Technology*, 147(2000) 158-164.

Section 4: Medical/Microbial
Paper presented at the Seventh International Symposium on Olfaction and Electronic Noses, July 2000

189

Detection of Bacteria Causing Eye Infections using a Neural Network Based Electronic Nose System

P Boilot[a], E L Hines[a], S John[a], J Mitchell[c], F Lopez[b], J W Gardner[a], E Llobet[b], M Hero[d], C Fink[c] and M A Gongora[a]

[a] Electrical & Electronic Engineering Division, School of Engineering, University of Warwick, Coventry CV4 7AL, UK
[b] Department of Electronic Engineering, Universitat Rovira i Virgili, Autovia de Salou s/n, 43006 Tarragona, Spain
[c] Micropathology Ltd, University of Warwick Science Park, Barclays Venture Centre, Coventry CV4 7EZ, UK
[d] Coventry and Warwickshire Hospital, Stoney Stanton Lane, Coventry CV1 4FH, UK

Abstract. An electronic nose (e-nose) data logger (Cyrano Sciences Inc, USA), comprising an array of thirty-two polymer carbon black composite sensors has been used to identify six species of bacteria, commonly associated with eye infections, over a range of concentrations in saline solutions. Readings were taken from the headspace of the samples by manually introducing the portable e-nose system into sterile glass vials containing a fixed volume of bacteria in suspension. After some data pre-processing, principal components analysis (PCA) and other exploratory techniques were used to investigate the clustering of the response vectors in multi-sensor space. Then, three supervised predictive classifiers, namely multi-layer perceptron (MLP), radial basis function (RBF), and Fuzzy ARTMAP, were used to identify the different bacteria. The optimal MLP network was found to classify correctly 97.3% of unknown bacteria types. The optimal RBF and Fuzzy ARTMAP algorithms were able to predict unknown bacteria with accuracies of 96.3% and 86.1%, respectively. A RBF network was able to discriminate between the six bacteria species even in the lowest state of concentration with 92.8% accuracy. These results show the potential application of neural network-based e-noses for rapid screening and early detection of bacteria causing eye infections and the possible development of a Cyrano e-nose as a near-patient tool in primary medical care.

1. Introduction

Despite the apparent robustness of the eye, there is no doubt that it is exposed to a harsh environment where it is continually brought into contact with pathogenic airborne micro-organisms. The function of the eyelids and production of tears help to protect the eye. However the warm, moist, enclosed environment, which exists between the surface of the eye (conjunctiva) and the eyelids, also provides an environment in which contaminating bacteria can establish an infection. The most common bacterial eye infection is conjunctivitis and organisms such as *Staphylococcus aureus, Haemophilus influenzae, Streptococcus pneumoniae, Escherichia coli* have been associated with this condition [1]. The number of organisms responsible for infection of the eye is relatively small; nevertheless the consequences are potentially serious because the eye may become irreversibly damaged. Rapid diagnosis is therefore essential but currently relies on the time-consuming isolation, culture and analysis of the infectious agent (e.g. optical microscopy). Since it is very important that the nature of the infection is diagnosed as quickly as possible, it is clear that techniques, such as neural network based e-nose that can almost instantly detect and classify odorous volatile components, would constitute a major advance in this area.

After nearly twenty years of development, e-nose technology has been applied in various industries such as the food, drinks and cosmetics. More recently, research has been directed towards health and safety issues [2], for example in medicine diagnosis, food quality and control, environmental monitoring. E-nose systems have already been used with success in the medical domain [3], for microbial detection [4], and bioprocess monitoring [5]. In this paper we shall describe the use of a prototype of the Cyrano Sciences e-nose to identify six species of bacteria, which are commonly associated with eye infections, when present at a range of concentrations in saline solutions. Exploratory data analysis techniques including PCA, fuzzy c-means (FCM) and self-organising maps (SOM) are used to investigate how the data cluster in multi-sensor space. Then, pattern recognition engines including MLP and RBF neural networks and the Fuzzy ARTMAP paradigm [6] are used for the classification of aromas according to the bacteria species and dilution states.

2. Experimental procedures

2.1. Materials

The bacterial samples used in this experiment are among the most common bacterial pathogens responsible for eye infection i.e. *Staphylococcus aureus* (sar), *Haemophilus influenzae* (hai), *Streptococcus pneumoniae* (stp), *Escherichia coli* (eco), *Pseudomonas aeruginosa* (psa) and *Moraxella catarrhalis* (moc). Tests were conducted at Micropathology Ltd., a medical laboratory based on the Sciences Park at Warwick, under typical laboratory conditions. All bacteria were grown on blood or lysed blood agar in standard petri dishes at 37°C in an humidified atmosphere of 5% CO_2 in air. After overnight culture, the bacteria were suspended in sterile saline solution (0.15M NaCl) to a concentration of approximately 10^8 colony forming units (cfu)/ml. A ten-fold dilution series of bacteria in saline was prepared and three dilutions (d1≡10^8, d2≡10^5 and d3≡10^4 cfu/ml) were tested using the e-nose. The numbers of viable bacteria present were confirmed by plating out a small aliquot of the diluted samples and counting the resultant colonies after incubation overnight.

2.2. Test procedures

The e-nose used was a prototype of the new Cyranose 320 (Cyrano Sciences Inc, USA), a portable system, whose component technology consists of 32 individual carbon black polymer composite sensors configured into an array. When the sensors are exposed to vapours (or aromatic volatile compounds) they swell, changing the conductivity of the carbon pathways and causing an increase in the resistance value that is monitored as the sensor signal. The resistance changes across the array are captured as a digital pattern that is representative of the test smell. The sensor technology yields a distinct response signature for each vapour regardless of its complexity; the overall response to a particular sample produces a 'smellprint' specific to a stimulus [7].

Data were gathered as follows. In each test, to collect samples the Cyrano data logger was introduced manually to a sterile glass vial containing a fixed volume of bacteria in suspension (4 ml). The operation was repeated ten times for each one of the three dilutions of each of the six bacteria species, to give a total of 180 readings.

3. Data analysis

3.1. *Signal pre-processing*

The choice of the data pre-processing algorithm has been shown to affect the performance of the pattern recognition stage. Software written in MATLAB (The MathWorks Inc, USA [8]) was used to extract the key features from the data in terms of the static change in sensor resistance, using a fractional difference model: $\Delta R = (R - R_o)/ R_o$ where R is the response of the system to the sample gas, and R_o is the baseline reading, the reference gas being the ambient room air. The complete bacteria data set was then normalised, by dividing each ΔR by the maximum value for each sensor, in order to set the range of each sensor parameter to [0,1]. This is necessary when Fuzzy ARTMAP is used, and for consistency, the same sensor normalisation technique was used for PCA, FCM, SOM, MLP and RBF. This also places equal weighting on the value of each sensor and so reduces bias. All exploratory and classification techniques were simulated and developed using MATLAB 5.

3.2. *Exploratory techniques and data clustering*

The use of PCA, FCM and SOM to assess clustering within the data set is now discussed. These exploratory techniques are used to examine the data clustering in multi-sensor space. Several techniques were applied to verify that the clusters established were genuine and the clusters formed matched the six types of bacteria. The objective of this step was to establish simple classes for the different bacteria species and see whether the data clusters could be found; before the pattern recognition stage.

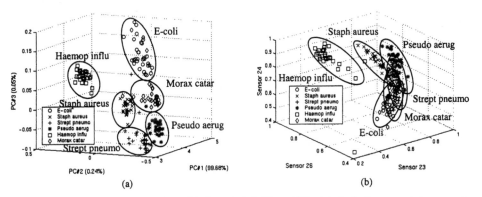

Figure 1: Clustering representations of the bacteria data; (a) on a PCA plot, six clusters appear to be apparent, and (b) on a 3D scatter plot using the least correlated sensors, categories appear to be less evident.

PCA is an effective linear projection method for discriminating between the e-nose responses to simple and complex odours [9]. The method consists of expressing the response vectors in terms of a linear combination of orthogonal vectors that account for a certain amount of variance in the data. The results of the PCA are shown in Figure 1(a), all concentrations are considered. Three principal components were kept, which accounted for 99.96% of the variance (PC 1, PC 2 and PC 3 representing respectively 99.68%, 0.24% and 0.05%), and 6 categories appear to be evident. The six clusters formed match the six types of bacteria so that the bacteria were completely separated in the principal component space with 94% accuracy. Most of the variance in the data is explained by considering only the first principal component, which implies that the sensor responses are highly correlated. Previous tests have shown that this is generally the case when using the data logger system, which suggests that some of the sensors could be omitted for future data analysis. In Figure 1(b), a 3-dimensional scatter plot of these data is given using the three least correlated sensors (23, 24 and 26); groupings appear to be less evident, so more than three are needed.

SOM and FCM were applied to the data-set in order to investigate clustering using the responses from the 32 sensors. A SOM network is a non-linear ANN paradigm, which is able to accumulate statistical information about data with no other supplementary information than that provided by the sensors [10]. Various SOM networks were created and trained with the entire data set, subsequently samples are associated with one of the neurones and neurones are grouped together to form categories of each identified bacteria. A 6x3 map achieved only 80% of correct clustering, whereas a 6x6 SOM network gave 90% accuracy. FCM is a fuzzy data clustering and partitioning algorithm in which each data point belongs to a cluster according to its degree of membership [11]. With FCM, an initial estimate of the number of clusters is needed so that the data set is split into c fuzzy groups. A cluster centre is found for each group minimising a dissimilarity function. The number of clusters was set to 6 and 18 to achieve 64% and 83% of correct clustering respectively. The high accuracy of the PCA results compare to the other two techniques suggests that the data could be easily linearly separated, as an investigation using pattern recognition methods now demonstrates.

3.3. Classification techniques and pattern recognition

The data-set was analysed using MLP, RBF and Fuzzy ARTMAP. The MLP network used is a standard feed-forward network trained with supervised back-propagation learning algorithm using either gradient descent with momentum (BPGDM) or Levenberg-Marquardt (BPLevMar) variations [12]. RBF networks are supervised learning paradigms very similar to MLP except that they use radial basis transfer function to classify data using hyper-spheres rather than hyper-planes [13]. Self-organising and self-stabilising Fuzzy ARTMAP generally performs better than does MLP, accumulating knowledge using associative memory to adjust the number of internal recognition categories, exhibiting fast learning for rare events [14]. Some earlier work [15] showed that it is a promising technique for e-nose data analysis.

For this analysis, the data, 180 response vectors from the test on bacteria, were divided into three folds of 60 vectors so that none of the vectors are repeated. The folds were formed so that each bacteria species and dilutions are distributed evenly in all of the three in proportion to its number of replicates. For training two of the folds are selected, 120 training vectors, and the last fold is used for testing. The folds are formed as follow:

Fold #1: 3 eco d1, 3 eco d2, 4 eco d3, 3-3-4 sar, 3-3-4 hai, 3-3-4 stp, 3-3-4 psa, 3-3-4 moc
Fold #2: 3 eco d1, 4 eco d2, 3 eco d3, 3-4-3 sar, 3-4-3 hai, 3-4-3 stp, 3-4-3 psa, 3-4-3 moc
Fold #3: 4 eco d1, 3 eco d2, 3 eco d3, 4-3-3 sar, 4-3-3 hai, 4-3-3 stp, 4-3-3 psa, 4-3-3 moc

3.4. *Prediction of bacteria species*

For this analysis, each of the folds contained ten replicates of each bacteria species irrespective of the level of dilution. The effect of dilution will be considered in section 3.5. Tests were repeated 5 times using each one of the three possible combinations of folds (2 to train and 1 to test) and extracting the mean for the summary results presented into Table 1.

ANN type	Topology	Data set	Accuracy %	ANN type	Parameter	Parameter	Accuracy %
MLP BPGDM	32-18-6	120tr-60tt	96.4	**RBF** 120tr-60tt	32 i/ps	spread 0.1	72.2
	32-18-6	480tr*-60tt	94.3		32 i/ps	spread 10	92.2
	8-8-6	120tr-60tt	75.5		32 i/ps	spread 100	96.3
	8-8-6	480tr*-60tt	78.9		8i/ps	spread 10	89.3
MLP BPLevMar	32-18-6	120tr-60tt	95.1		8i/ps	spread 100	93.8
	32-18-6	480tr*-60tt	97.3	**Fuzzy** **ARTMAP** 120tr-60tt	ρ_a=1.0	120 categ	85.5
	8-8-6	120tr-60tt	92.9		ρ_a=0.99	85 categ	86.1
	8-8-6	480tr*-60tt	94.1		ρ_a=0.95	24 categ	69

* 480 training vectors were generated from the 120 train vectors (2 train folds) using 5% of random noise.

Table 1: Results of the type of bacteria classification with MLP train with BPGDM and BPLevMar, RBF and Fuzzy ARTMAP, in terms of correct prediction percentage.

- For MLP, the 32-18-6 network used had 32 inputs, 18 hidden nodes and 6 output neurones, since a one-of-six target was used to code the six different bacteria species. Both *back-propagation* networks used the same architecture, however the one trained with BPGDM had a *learning rate* set to 0.05, a *momentum term* to 0.9 and 20,000 *epochs*, the other one was trained with BPLevMar. A smaller network (8-6-6) was also trained using the 8 least correlated sensors as inputs but in general, it proved to be less efficient. The use of additive random noise to generate more training vectors and thus potentially improving the generalisation of the network was also investigated but when using folds for training it has no significant influence on the results.
- For RBF, the model implemented under Matlab used the so called *more efficient design* in which RBF neurones are added one at a time until the sum-squared error falls beneath the *error goal* (set to 10^{-5}). A few values for the *spread* constant, which determines the sensitivity of the input range and overlapping, were tested. A value of 0.1 means over fitting whereas a value of 100 implies large overlapping regions. Here, again both 32 and 8 inputs networks were implemented, again 32 inputs were found to be necessary to achieve high classification rate.
- For Fuzzy ARTMAP, the data were analysed using Matlab code developed by us after having verified its functionality. The network has 32 inputs and 6 outputs and was trained only for one epoch. The value of the *choice parameter* was set to an optimal value of 0.001. The number of hidden units or categories varies according to the value of the *vigilance parameter* ρ_a in ART$_a$. The *recode parameter* β_a was set to 1 for fast learning and the network was trained with values of ρ_a set to 1, 0.99 and 0.95. With a value of 1, generalisation is poor as the numbers of categories formed is equal to the number of training patterns, no improvement is possible as there is only one epoch but then the classification rate is high. With a value of 0.95, the number of class formed is lower but then the correct classification rate is slightly lower too.

For the MLP BPGDM network the number of epochs was set to 20,000 and converged within less than 100 with BPLevMar, however, a long computational time is required for both systems, whereas for RBF or Fuzzy ARTMAP the time necessary to train is much smaller, less than one minute. The optimal MLP network was found to classify correctly 97.3% of unknown bacteria types. The optimal RBF and Fuzzy ARTMAP algorithms were able to predict the species of unknown bacteria with accuracies of 96.3% and 86.1%, respectively. The relatively poor classification rate for Fuzzy ARTMAP is due to the low results obtained when tested with Fold 1, which contains most of the rare events difficult to classify when unseen. High classification rates for both MLP and PCA show that the bacteria types are linearly separable.

3.5. Prediction of bacteria dilutions

For this analysis, different objectives were considered such as: creating a system able to separate for one bacterium each dilution state; creating a system to discriminate bacteria species in one specific dilution; and considering the overall problem using the same folds as the previous analysis to find both bacteria and dilution using two target coding techniques.

- The first approach considered used the same folds as before but then the data set was divided into 18 classes (6 bacteria x 3 dilutions). The MLP network trained had 18 outputs and achieved 25.4% of correct classification with BPGDM and 54% with BPLevMar. RBF and Fuzzy ARTMAP algorithms were able to predict the species and dilution of unknown bacteria with accuracies of 58.9% and 42.2%, respectively. The low results for this particular implementation are due to the extreme complexity of the systems formed and the relatively small amount of data vectors to correctly train them.
- In the second approach, the aim was to discriminate, for a given bacteria, the various dilution states. A 32 input RBF network was implemented for this analysis because previous tests show that it converges rapidly and gives accurate results. Only 30 vectors are available from each bacteria data set, 21 vectors selected randomly were considered for training (7 per dilution) and 9 for testing. Psa was the most difficult bacteria to separate with 46%, then 52% for hai, 66% for sar, 67% for moc, 68% for eco and finally 73% for stp. The relatively low percentage, an overall average of 62%, is mainly due to two things; the e-nose specifications, a system geared towards quick *smell and tell* applications and not designed to be quantitative but still accurate and reliable, and the limitations due to the sensor normalisation pre-processing technique, which tends to remove concentration information.
- In the third approach, one state of dilution is taken into account and a RBF network is developed to recognise the different bacteria species. Three data-sets are considered, one for each dilution, containing 60 vectors, 42 were randomly selected for training (7 per bacteria) and 18 for test. For d1, highest concentration 10^8 cfu/ml, the classification rate is 90.3%, 96% for d2 and 92.8% for d3, 10^4 cfu/ml lowest concentration. Even if the e-nose does not appear to be able to correctly discriminate concentration states for one given bacterium, it can be used to discriminate between bacteria species even at low concentration. The overall average accuracy for these tests is 93%.
- In the last approach, again folds were used, but this time only 6 outputs were considered, one per bacteria species (1 in 6 coding), a strength was given to the target output to represent the bacterial concentration. For d1, highest concentration, the target was set to 1, it was set to 0.7 for d2 and 0.5 for d3, all the other outputs being 0. MLP when trained with BPLevMar achieved 64.5% of correct classification whereas RBF gave 63.9%. These results show that it is still difficult to consider the problems of species and dilutions together within one classification system.

Nevertheless, these results on the dilution problem are encouraging in parts and further investigations should show that the Cyrano nose could be use for the discrimination of bacteria causing eye infections even at low concentration level. At 10^4 cfu/ml, our highest dilution, very close to real medical domain eye infection concentration, the system was able to discriminate between bacteria species with 92.8% accuracy.

4. Conclusions

Odour patterns from six different bacteria species believed to be responsible for eye infections in three dilution states have been analysed by an e-nose. First PCA, FCM and SOM were used as exploratory techniques to investigate clustering of the entire data set within the multi-sensor space according to the expected bacteria species. Then, MLP, RBF and Fuzzy ARTMAP neural networks were applied in order to classify the bacteria species using the entire data set, with no regards to the dilutions. An accuracy of 97.3% was reached in the classification using MLP, 96.3% was obtained with RBF and 86.1% was obtained using Fuzzy ARTMAP. It was found that the low rate for Fuzzy ARTMAP is due to the low results when fold 1 is used for testing. Fuzzy ARTMAP achieve a good 97.6% accuracy when fold 1, which contains unique data points, was used for training. All techniques seemed to reach similar classification rate on bacteria species comparable to traditional MLP networks. Furthermore, the network training times were found to be much faster with RBF and Fuzzy ARTMAP (few minutes) than those for MLP (hours), which make them more useful learning paradigms.

Finally, similar MLP, RBF and Fuzzy ARTMAP neural networks were applied to the problem of predicting bacteria species along with dilution states. It was found that when using RBF networks, the data were more difficult to classify according to the dilution states when considering one specific bacterium (62%) than they were to classify according to the bacteria species when using data sets for one particular concentration (93%). This analysis was carried out in order to investigate the effect of concentrations, even if the e-nose did not perform well on quantitative discrimination of dilution states, nevertheless, it was able to discriminate between the six bacteria species even in the lowest state of concentration with 92.8% accuracy. Another approach dealt with the entire data-set, a 1 in 6 coding was used to discriminate bacteria type, then a relative strength was given to the target output considering the dilution state, 1 for highest d1, 0.7 for d2 and 0.5 for lowest d3. An accuracy of 64.5% was reached using MLP BPLevMar and 63.9% was obtained with a RBF network. The MLP model was found to perform slightly better than RBF but took a lot longer to be trained.

All these results make the RBF network very attractive for pattern classification in the context of real e-nose diagnosis tool for eye infections. This application of the Cyrano e-nose device shows that this technology is capable of distinguishing bacteria responsible for eye infections even at low concentration states, 10^4 cfu/ml close to eye infection concentration. The sensor technology employed does not require precise sampling methods and/or heavy experimental protocol to achieve full separation of the pathogens of interest. However, further work is needed to assess the long-term reliability of the system, in particular the effects of both sensor drift and concentration level on detection rates. The next step is to analyse the air taken directly off a patient's eye, however the odours emitted by the human body will have to be taken into account in order to create a reliable model. Nevertheless, we believe that a neural network-based portable electronic nose, such as a modified version of the Cyranose, may provide an attractive near patient diagnostic tool for rapid screening, and thus potentially provide better patient healthcare for the treatment of eye infections.

Acknowledgments

We thank Micropathology Ltd. for the use of laboratory and assistance when running the tests, Cyrano Sciences for the e-nose system and technical assistance. We also thank Felix Lopez for his active collaboration and helpful support in processing the data. Pascal Boilot gratefully acknowledges financial support (award number 99310943) from EPSRC during his stay as a PhD student at the University of Warwick.

References

[1]   *Infections of the eye*, 1993 Medical Microbiology eds Mins et al. 21 1-4
[2]   Gardner J W and Bartlett P N 1999 *Electronic noses: principles and applications* Oxford University Press
[3]   Di Natale C, Mantini A, Macagnano A, Antuzzi D, Paolesse R and D'Amico A 1999 *Electronic nose analysis of urine samples containing blood* Physiol. Meas. 20 377-84
[4]   Gardner J W, Craven M, Dow C S and Hines E L 1998 *The prediction of bacteria type and culture growth phase by an electronic nose with a multi-layer perceptron network* Meas. Sci. Technol. 9 120-7
[5]   Shin H W, Llobet E, Gardner J W, Hines E L and Dow C S 2000 *Classification of the strain and growth phase of cyanobacteria in potable water using an electronic nose system* IEE Proc. - Sci. Meas. Technol. 147 158-64
[6]   Llobet E, Hines E L, Gardner J W and Franco S 1999 *Non-destructive banana ripeness determination using a neural network-base electronic nose* Meas. Sci. Technol. 10 538-48
[7]   www.cyranosciences.com
[8]   www.mathworks.com
[9]   Gardner J W 1991 *Detection of vapours and odours from a multi-sensor array using pattern recognition, part 1: principal components and cluster analysis* Sensors Actuators B 4 108-16
[10]   Kohonen T 1987 *Self-organising and associative memory* 2nd edition Berlin: Springer-Verlag
[11]   Jang J S R, Sun C T and Mizutani 1997 *Neuro-fuzzy and soft computing: a computational approach to learning and machine intelligence* Upper Saddle River NJ: Prenctice Hall 423-33
[12]   Hagan M T, Demuth H B and Beale M H 1996 *Neural Network Design* PWS Publishing Company Boston MA Chap 11 and 12
[13]   Chen S, Cowan C F N and Grant P N 1991 *Orthogonal least squares learning algorithm for radial basis function networks* IEEE Trans. on Neural Networks 2 302-9
[14]   Carpenter G, Grossberg S, Markuzon N, Reynolds J and Rosen D 1992 *Fuzzy ARTMAP: a neural network architecture for incremental supervised learning of multidimensional maps* IEEE Trans. On Neural Networks 3 698-713
[15]   Llobet E, Hines E L, Gardner J W, Bartlett P N and Mottram T T 1999 *Fuzzy ARTMAP based electronic nose data analysis* Sensors Actuators B 61 183-90

Detection of Dry Rot in Timbers using a Hand-held Electronic Nose

P D Wareham, H Chueh[1], K C Persaud, P A Payne, J V Hatfield[1]

Department of Instrumentation & Analytical Science, and [1]Department of Electrical Engineering & Electronics, UMIST, PO Box 88, Manchester, M60 1QD, UK

Abstract. The fungal attack of timber by *Serpula lacrymans*, resulting in the occurrence of dry rot, causes extensive damage to susceptible buildings. Remedial treatments are often costly since early detection of such fungal attack is problematic. *Serpula lacrymans* produces a distinctive odour that has been shown to comprise of defined metabolic volatiles. Such a volatile fingerprint could be used as a marker for infection using a portable electronic nose system. This paper describes an approach towards the development of such a device for early on-site detection of the presence of dry rot in timbers.

1. Introduction

'Dry rot' is the term used to describe the result of an outbreak of the wood decay fungus *Serpula lacrymans* (*S. lacrymans*), an attack of which can cause extensive damage to susceptible building timbers. Fungal attack results in reduction of the structural integrity of the host timber, which if left untreated, can progress to a stage where timber readily crumbles with resultant compromise to building safety. Remedial treatments are elaborate, involving the eradication of moisture sources, promotion of rapid drying, replacement of decayed timbers and application of fungicide to remaining sound timbers. Such countermeasures are often costly and where treatments are impractical, or the extent of damage is significant, building demolition is sometimes economically preferable to rehabilitation. The specific growth conditions required by *S. lacrymans* are readily found in areas such as cellars, behind panelling and plaster and in sub-floor spaces, which makes early detection of the fungus visually problematic, despite a defined physical appearance. An alternative detection method would therefore prove invaluable to aid the early diagnosis of *S. lacrymans* attack, both minimizing costs associated with fungal damage and disruption caused by remedial investigations.

The potential of electronic nose technology for detection of wood rotting fungi has previously been demonstrated for fungal infections causing brown and soft rots [1]. A distinctive mushroom-like odour accompanies *S. lacrymans* growth that also provides potential for use as a marker in such a system. The volatile composition of this odour has been investigated in a number of studies, which suggest the presence of a range of metabolic volatile organic compounds (MVOC) [2-4].

With the possibility of defined MVOC fingerprints being attributable to *S. lacrymans* developing in various conditions, there is scope to develop a portable electronic nose system capable of dry rot detection. Such a device would provide advantages such as immediate on-site analysis and non-invasive sampling protocols. An approach towards the development of such a device is presented here.

2. Development

Ongoing research and development is focussed in a number of areas:

2.1. Fungus culture and MVOC analysis

There is a need to eliminate any possible interference to a MVOC analysis from external factors such as sample contamination. A fungus culture system has been developed, using a hybrid method based on the work of Korpi et al and Low et al, to provide a reliable sample source [4,5]. Sterile 'microcosms' are used to provide suitable conditions of temperature and humidity to enable fungus culture on agar plates prior to inoculation of timber samples.

A comprehensive study of MVOC's originating from a range of wood rotting fungi has been undertaken by Esser and Tas [3]. Table 1 summarizes those detected by gas chromatography mass spectrometry (GCMS) for *S. lacrymans* developing on malt agar in two growth stages. A similar study by Korpi et al [4] for *S. lacrymans* growing on aspen suggests positive and negative indicators associated with fungal infection.

Table 1. Detected MVOC's from *S. lacrymans* as reported by Esser and Tas [3].

	Detected MVOC	
m/z (M$^+$)	*S. lacrymans* (vegative)	*S. lacrymans* (sporulating)
72	2-butanone	2-butanone / tetrahydrofurane
74	1-butanol	
82	2- or 3-methyl-furane	
84		dichloromethane / methylcyclopentane
86	2-methyl-3-butene-2-ol / 3-methyl-2-butanone	C_6H_{14}*
94	dimethyldisulfide*	
96	furfural	
100		C_7H_{16}
106		benzaldehyde
112	5-heptene-2-on*	
126	furanecarbonacid methylester	
128	2,2-dimethyl-hexanal* / branched ketone*	
132		1,1,1-trichloroethane
166		tetrachloroethene

*Compounds were detected in significant quantities.

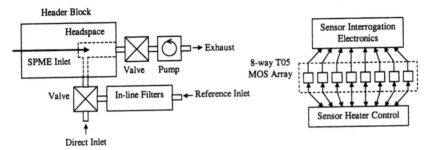

Figure 1. Schematic overview of measurement system.

Positive indicators comprise of formaldehyde, acetone, 2-pentanone and α-pinene, with negative indicators comprising of acetaldehyde, 1-hexanol, hexanal and nonanal, in descending orders of abundance respectively.

Comparison of such data suggests that MVOC emissions are likely to be both substrate and growth stage dependent. It is important, therefore, to undertake similar analysis of *S. lacrymans* in a range of controlled environments to enable the appropriate selection of sensor systems for effective detection of dry rot.

2.2. Sensor systems and sampling protocols

Korpi et al [4] reports concentration ranges of *S. lacrymans* MVOC's of the order of 2 to 200 ppb, which suggests that a headspace sampling protocol may require a pre-concentration step to enable optimal sensor response to be achieved. The use of metal oxide semiconductor (MOS) sensors in combination with solid phase micro extraction (SPME) headspace sampling is being investigated for this purpose.

SPME sampling and MOS sensors are complimentary since rapid desorption of trapped MVOC's from the SPME fibre into a sensor headspace can be stimulated by the sensor heaters. Moreover, SPME provides scope for selective MVOC adsorption to reduce sample noise from ambient odour backgrounds and de-coupling of humidity variation to improve sensor stability. SPME also provides advantages in terms of ease of use and applicability to field operation.

A schematic overview of the prototype electronic nose system, developed to enable the use of commercial SPME and MOS devices, is given by figure 1. An 8-way MOS sensor array, temperature and humidity probes are housed in a stainless steel header block assembly. The valves and pump enable headspace isolation for passive thermal desorption of SPME sampled volatiles, or direct headspace sampling using a flow-through approach.

2.3. Instrumentation

The prototype incorporates a microprocessor-based interface that measures sensor resistance change and then transfers data to a host PC for more advanced data analysis. Figure 2 outlines system design, based on a micro-controller sensor interface and hand-held PC.

The interface design incorporates self-calibration of the electronics and sensor interrogation through the following steps; sequentially select sensors within the array via a multiplexer, program a suitable voltage for the resistance interrogation circuit, cancel the input offset voltage from the operational amplifiers and the on-resistance from the multiplexer channels, balance a large range of baseline resistances (from hundreds of Ohms to tens of mega Ohms) and detect small sensor response resistances.

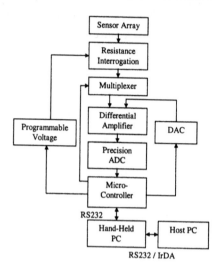

Figure 2. Schematic overview of the sensor interrogation interface.

Communication between the interface and hand-held PC occurs via an RS232 serial communication protocol. Custom software enables collection and storage of data and provides immediate on-site analysis. It also permits remote control (via a wire communication interface or an infra-red communication interface) allowing a host PC to import stored data for further analysis.

3. Discussion

Detection of the presence of *S. lacrymans* in timber using a portable electronic nose has the potential to provide a valuable on-site aid for early diagnosis of dry rot. Such a device could vastly reduce the costs associated with extensive remedial treatments and provide an alternative, non-invasive investigative strategy. The proposals and developments outlined in this paper constitute ongoing research and it is expected that significant advances towards a hand-held device will be made in the future.

References

[1]   Nilsson K 1998 PhD Thesis (Swedish University of Agricultural Sciences)
[2]   Bjurman J and Kristensson J 1992 International Research Group on Wood Preservation. Doc. No. IRG/WP/2393-92
[3]   Esser P M and Tas A C 1992 International Research Group on Wood Preservation Doc. No. IRG/WP/2399-92
[4]   Korpi A, Pasanen A and Viitanen H 1999 Building and Environment 34 205-11
[5]   Low G A, Palfreyman, J W, White N A and Sinclair D C A 1999 Holzforschung 53 129-36

Section 4: Medical/Microbial
Paper presented at the Seventh International Symposium on Olfaction and Electronic Noses, July 2000

201

Techniques to Allow the Detection of Oestrus in Dairy Cows with an Electronic Nose

T. T. Mottram, R.M. Lark, A.J.P. Lane†, D.C. Wathes†, K.C. Persaud‡, M. Swan, J.M. Cooper•

Silsoe Research Institute, BEDFORD, MK45 4HS
Phone: +44 1525 860000, Fax: +44 1525 861735, E-mail:
toby.mottram@bbsrc.ac.uk.
†Royal Veterinary College, Potters Bar, Hertfordshire.
‡DIAS, UMIST, Sackville St, Manchester.
• Nanotechnology Centre, University of Glasgow.

Abstract. A portable electronic nose was constructed to detect the odours released by cattle in oestrus. It consisted of conducting polymer sensors, a system for sampling, humidity management (known as the olfactory lens) and data analysis. It was used in a number of experiments to detect oestrus in cows with different polymer based electronic noses. The method developed was to take swabs of cervical mucus from the cows and place them in a sample chamber. The humidity emanating from the swab was measured and the humidity surrounding the sensors was adjusted to minimise the change in humidity experienced by the sensors. The change in electrical signal measured by the sensors was recorded. A novel method of data analysis – the wavelet function was adapted and applied to the transient signals from the odour sensors. The wavelet function coefficients characterised the signal from each sensor. By taking the ratio between the coefficients representing the most active sensors in the array a two dimensional vector plot of each days sample was drawn. The method was applied to four cows of whom three showed a clear change in odour found in their mucus samples between the day of oestrus and the day of ovulation. This project has shown that with improvements in sample management (the olfactory lens) and signal processing (wavelet analysis) existing electronic noses can be used to detect oestrus in dairy cows but with an invasive sampling technique.

1. Introduction

The improved detection of oestrus in dairy cattle is a major objective in livestock engineering. The onset of oestrus is signalled in dairy cattle by pheromones emanating from the perineal glands on either side of the vulva (Blazquez et al 1988) and present in cervico-vaginal mucus (Klemm et al 1987). Odours from the cow have been demonstrated to be the signal to the bull that the cow is receptive for natural mating (Kiddy, 1984). The compounds have yet to be reproducibly identified so an odour sensing system would be appropriate for non-invasive detection of oestrus in lactating and non-lactating animals. The first problem to resolve is to determine what samples can be reproducibly taken. A pilot study was conducted with an AlphaMOS electronic nose to determine the relative strength of signals from swabs taken at sites on 5 cows during the mid luteal phase and at oestrus. The 3 regions were the perineal region, the cervical os and the mouth (control). The AlphaMOS nose used metal oxide sensors that are less susceptible than conducting polymers to humidity effects. Nevertheless the differences in the amplitude of signals measured with the eight sensors clearly divided the samples into wet and dry. These differences were greater than those attributable to stage of cycle. However, it was possible to distinguish oestrus from luteal samples in the perineal swabs using principal components analysis.

In a separate study swabs from the perineal areas of 8 cows were tested over several days with a Neotronics "NOSE". Of the 12 sensors tested, 7 showed a significant change in resistance that was dependent on the day of the oestrous cycle and was highly correlated with the plasma oestradiol concentration (Lane and Wathes 1998). However, the perineal site is problematic for sampling in dairy cows as it is frequently contaminated with faeces. Although the collection of cervico-vaginal mucus is a more invasive process it offered the advantage of avoiding faecal contamination.

As humidity was a major factor affecting the conducting polymers, studies were conducted to determine how to minimise this. Two conventional humidity generators (General Eastern) were set up to create flows of air at known humidity. The air streams were passed over two sensors provided by Glasgow University. When the air streams were switched from one humidity level to another the change in output was recorded. The results indicated that a calibration curve for individual sensors can be derived in this way. The response was linear but with different slopes for two conducting polymer sensors ($r^2=0.89$ and $r^2=0.96$) across the range of 40 to 80 % relative humidity.

The common method of analysing electronic nose data was to take a single point from a signal response versus time plot, typically this is the amplitude at some fixed point in time. After the data point had been taken the data that related to rate of absorption and desorption of the odour into and from the sensor were discarded. The single data points would then be put into a pattern recognition procedure such as principal components analysis that would transform the n dimensional data (n being the number of sensors in the array) into a two or three dimensional plot. Unfortunately such an approach hides the physical reality of the response of the sensors. The information in a trace from an electronic nose sensor may be static or transient and thus a more appropriate pre-processing algorithm the discrete wavelet transform (DWT) is here proposed to characterise the transient and static information in a signal from an electronic nose.

The objectives of this work were to show that electronic noses can detect oestrus in dairy cows and that this can be achieved by controlling humidity as samples are presented and that the wavelet method of pre-processing data from sensors can simplify the electronic nose data analysis.

2. Materials and Methods

A device known as the olfactory lens was built to give a constant flow of air with programmable variable relative humidity air flow into a sample chamber such that a maximum change of humidity (from 0 to 90 % rH) could be achieved in 30 s. A system was required that could pre-condition the sensors to the relative humidity emanating from the sample. A change in signal from the sensors when confronted with odours from the swab would then signify a change in odour rather than a change in humidity.

A system was designed and built (Mottram et al, 1999) to control the flow of air over a sensor. A swab could be placed in the sample chamber aspirated by a known volume of air within a closed loop of pipe. A humidity probe measured the relative humidity that the swab induced within the closed loop. The humidity of the flow of clean air passing over the sensors was then adjusted to match the humidity found in the sensing chamber. The flow circuits were then switched so that a flow of odorous air from the swab passed over the sensors.

In Experiment 1 to use the olfactory lens 8 ovariectomised cows were brought into oestrus experimentally and any signs of oestrous behaviour were noted. Duplicate cervical mucus swabs were taken, the first samples pre-treatment with oestradiol benzoate, and then at 12 hour intervals over the following three days. The swabs were frozen at -80°C until subjected to odour analyis and GC-MS.

Immediately before odour analysis, each frozen swab was thawed for 5 minutes in a glass bijou bottle at 25°C. The bottle was connected to the inlet of the Aromascan A32S electronic nose, and odourants stripped from the sample by a small pump and drawn over the sensor array for 5 minutes. The background level of humidity was held at 70% rH, this level gave a small positive step change in humidity when the electronic nose sensors were exposed to the sample. Samples from each cow were read in a random order, and the sensor chamber was flushed with background air at 70% rH between samples to prevent carryover of odourants. Samples of water were read between each batch of 7 samples from a cow.

The effect of operating the olfactory lens was to reduce the amplitude of the change in resistance of the sensors by a factor of 10. However, this still gave full scale deflections in the order of 0.1 % change of resistance and analysis of the signal is still possible and gives useful results (Figure 1).

Experiment Two was conducted to test the potential for using wavelet analysis to simplify preprocessing of data. Daily samples of cervical mucus were taken. The oestrus cycle was mapped by frequent analysis of milk for progesterone and plasma oestradiol concentrations. The time of ovulation was determined by ultrasound scanning as described previously (Lane and Wathes 1998). Four cows were studied in this way for 30 days each.

Figure 1. Results from removing the humidity effect from an Aromascan nose used to analyse swabs from an ovariectomised cow treated with oestradiol benzoate (0.5 mg im). One sensor from the array could be selected and used as a specific oestrus odour detection system giving an output in engineering units.

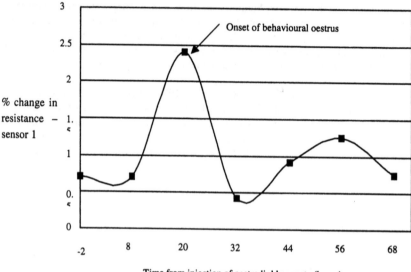

Time from injection of oestradiol benzoate (hours)

A modified version of the olfactory lens was used to precondition the sensor chamber to the mean humidity emitted by the cervical swabs (dimensional irregularities of the sensors prevented the full implementation of the olfactory lens). The odours from the swabs were passed over the sensors for 512 s. The change in resistance of each sensor was divided by the base resistance of the sensor at the start of the individual sample analysis.

2.1   Selection of sensors to use with wavelet analysis.

The nose used had eight sensors. The correlation matrices for the eight sensors were examined for two of the cows from Experiment 2 on two days, corresponding to mid-luteal conditions and the day of ovulation. This showed that sensors 1 to 6 and 8 highly intercorrelated (most correlation coefficients of 0.95 or more). Sensor 7 seemed to be distinct from the rest with correlation coefficients with the other sensors between 0.43 and 0.8 on the mid-luteal days and between -0.22 and 0.06 on the day of ovulation. Sensors 1 and 7 were therefore selected for more detailed analysis.

The data consisted of measurements taken at a frequency of one per second. The result of the analysis was 510 wavelet coefficients and 2 mother function coefficients for each sensor on each day of the experiment. The analysis was limited to the full week up to and including the day of ovulation, this being the longest sequence available for all cows.

2.2   Identifying combinations of wavelet coefficients for detecting oestrus

The mother-function coefficients (mfc) were first examined.

While the mfcs appear to change in the days up to ovulation, there was not a consistent pattern. Attention was therefore focused on the transient information assumed to be present in the wavelet coefficients.

The square of a wavelet coefficient is an additive component of the sum-of-squares of the data set from which it is derived. By examining the squared wavelet coefficients for all cows on the sixth day before ovulation and the day of ovulation, it was seen that substantial contributions to the variance of the trace were made by components at scales 256, 128 and 64 seconds (i.e. the three coarsest-resolution components of the trace). Further work was therefore focused on these.

In addition the products of the wavelet coefficients for sensors 1 and 7 at the coarsest scale were computed. As the squared wavelet coefficients are components of the sum-of-squares of the data, so the products of the corresponding coefficients are components of the sum-of-products of two data sets. It was decided to investigate the products of the wavelet coefficients at this broadest scale of resolution, since the two sets of correlation matrices investigated in section 2.1 above both showed a difference in the correlation of sensors 1 and 7 between the mid-luteal trace and the trace on the day of ovulation.

Canonical variate analysis was used to explore how the wavelet coefficients might identify oestrus and ovulation. This procedure identifies a linear combination of a set of N variables which defines one or more new quantities (the canonical variates) which most effectively distinguish between g groups of observations. c canonical variates may be defined where c is equal to the smaller value of N and $(g-1)$. In this case two groups were defined: electronic nose traces corresponding to days when oestrous signs were observed OR ovulation took place, and traces corresponding to days with neither ovulation nor oestrous signs (luteal or pre-oestrus days). A single canonical variate was therefore defined. It was decided to carry out the analyses separately for the six wavelet coefficients for scales 256 and 128 s and for the eight wavelet coefficients at scale 64 s. This reduces the size of the covariance matrices which are being estimated from very limited data.

Canonical variates analysis was carried out, as described above, for the wavelet coefficients of the data for sensors 1 and 7 of the electronic nose, and for the products of the wavelet coefficients at the same scales. Given the limited number of observations available, and the fact that those observations made on separate days from the same cow cannot be regarded as independent of each other, this analysis should be interpreted cautiously. This transformed the data on several variables to the space in which the different states of the cows are separated as clearly as possible.

3   Results

Experiment 1. An odour trace was obtained from all of the samples. The small humidity offset between background and samples gave a rapid increase in signal immediately on exposure of the sensors to the sample, due to the water molecules present, followed by a more gradual change in signal as odourant molecules from the sample were absorbed into the sensors. The humidity signal allowed the odour traces to be separated from the baseline, which in turn enabled consistent analysis of the traces with the proprietary software package of the Aromascan electronic nose. The software allows a window of time in the odour trace produced by plotting the resistance of the 32 sensors of the nose against time to be isolated. The resistance data from a 30 second window at the end of the odour reading period were extracted. Principal component analysis enabled the multi-dimensional data from each odour sample to be represented as a point on a two- or three-dimensional graph, or 'odour map'.

Seven of the 8 cows came into oestrus on average 27h after the oestradiol benzoate injection (range 16-32h). Isolating the output of one sensor that showed a large difference in response over the course of the experiment and plotting the maximum response of this sensor against time showed a clear peak in response 8-20h after the start of treatment in 5 of the 8 cows (Figure 1). The results for 2 of the remaining 3 cows also showed a peak at 20h, but with a high initial reading. This may have been due to blood contamination of the swab from a vaginal biopsy sample collected just before swabbing. The remaining cow had a high initial reading followed by a low sensor reading with no clear peak.

Experiment 2.

Table 1 Wavelet coefficients for outputs of sensors 1 and 7. The data in row 3 indicate the significance of the difference between odour from cows on days when they were on heat.

Sensor	1		7		1,7	
Variables	wavelet coefficients		wavelet coefficients		product of wavelet coefficients	
Scales	256 128	64	256* 128	64*	256*	
loadings in canonical variate	-1.22 -3.42 -17.25 -0.86 -0.05 -7.67	-7.04 -3.46 20.75 3.56 -25.07 9.32 -4.13 -3.73	2.29 -2.76 0.58 4.54 2.15 8.87	11.81 -9.14 -3.41 1.47 -12.44 11.98 -2.94 1.06	1.055 2.838	
p Ho of no difference among groups	0.85	0.19	0.018	0.09	0.071	

*The loadings in these columns are multiplied by 10^3

Figure 2 A plot of wavelet coefficients 1 and 3 show a movement from the bottom left hand quadrant toward the top right hand quadrant. Three of four cows showed these patterns. This method of pre-processing may remove the need for pattern recognition software.

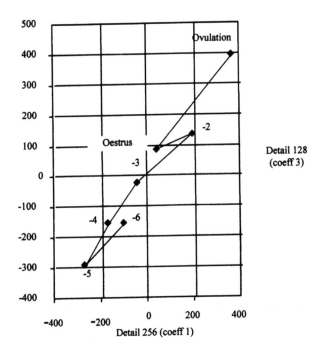

Detail 128 (coeff 3)

Detail 256 (coeff 1)

The strongest evidence for a difference between the traces for cows in oestrus or ovulating and the rest is seen in the wavelet coefficients at scales 256 and 128 for the output for sensor 7. The highest loadings are seen for the second and fourth wavelet coefficient at scale 128, but it is questionable if this small data set permits the deduction that these coefficients are the most important for discriminating oestrus and ovulation. The canonical variate generated by these loadings is plotted for the days before ovulation in Figure 2. There is also some evidence that the products of the wavelet coefficients at scale 256 for both sensors distinguish oestrus (when oestradiol levels would be high) and ovulation, (when oestradiol levels have fallen). Each point is labeled with the number of days before oestrus. It appears that a marked movement towards the upper left quadrant of this plot may indicate oestrus and/or ovulation. However, the limited data available can only be used to indicate that this is a promising approach to analysing the output of the electronic nose for the purpose of oestrous detection.

4. Discussion

These experiments have shown that with improvements in sample management (the olfactory lens) and signal processing (wavelet analysis) existing electronic noses can be used to detect oestrus in dairy cows. However, this application may be an inappropriate use of this technology. The electronic nose has its greatest advantage where the sole other means of measuring the signal is the human olfactory sense such as in food taste or odour detection (for example assessing grain quality). Where an identified compound is to be detected, then the novel technology of immunosensing would appear the most appropriate method of detection. An immunosensor to measure progesterone in milk would now appear to be the most appropriate method of predicting the time of ovulation.

Acknowledgements

The principal funding for this project was from the Ministry of Agriculture, Fisheries and Food, Great Westminster House, London. Prof. P. Bartlett, Southampton University lent us an AlphaMOS nose to commence the studies. We thank Miss L Hicking, University of Nottingham for assistance with the study on ovariectomised cows.

References

Blazquez N B French J M Long S E and Perry G C 1988 A pheromonal function for the perineal skin glands in the cow Veterinary Record 123 49-50

Kiddy C A Mitchell D S and Hawk H W 1984 Estrus-related odours in body fluids of dairy cows Journal of Dairy Science 67 388-391

Klemm W R Hawkins G N and De Los Santos 1987 Identification of compounds in bovine cervico-vaginal mucus extracts that evoke male sexual behaviour Chemical Senses 12 77-87

Lane A J P and Wathes D C 1998 An electronic nose to detect changes in perineal odours associated with oestrus in the cow Journal of Dairy Science 81 2145-2150

Mottram T T Vass S and Houghton C 1999 Apparatus and methods relating to humidified air and sensing components of gas or vapour Patent Application 9824556.6

Mottram T T and Frost A R 1997 Report on the state of the art of current and emerging methods of oestrus detection on UK Dairy Farms Report CR/775/97/1681 pub Silsoe Research Institute

Applications:
Food, Agricultural and Environmental

Section 5: Applications: Food, Agricultural and Environmental
Paper presented at the Seventh International Symposium on Olfaction and Electronic Noses, July 2000

211

A Dedicated Wheat Odour Quality Measurement System

Phillip Evans[†], Krishna C. Persaud[+*], Alexander S. McNeish[†],
Robert W. Sneath[‡], R. Norris Hobson[‡] and Naresh Magan[ᴴ]

[†] Osmetech plc, Electra House, Electra Way, Crewe, CW1 6WZ
[+]DIAS, UMIST, Chemistry Tower, Faraday Building, Sackville St. Manchester, M60 1QD[†]
[‡] Silsoe Research Institute, Wrest Park, Silsoe, Bedford, MK45 4HS
[ᴴ]Cranfield Biotechnology Centre, Cranfield University, Cranfield, Bedfordshire, MK43 0AL

* Corresponding author

Abstract. We report the outline construction and use of a dedicated sampler for the discrimination of wheat samples based on odour for at-line measurement giving real-time grading of samples. Commercial grain samples were classified as good or bad by a trained odour panel and the results compared against the dedicated sampler with promising results.

1 Introduction

We have designed and built a device for measuring the quality of wheat based upon its mouldy odour, the Grain Automated Sampler Prototype (GASP). It comprises an automated sample presentation system with a polymer based sensor array providing data for a Radial Basis Function artificial neural network (RBFann) capable of delivering real-time odour evaluation of the sample presented. The development of the Radial Basis Function artificial neural network (RBFann) for use in association with wheat odour measurements has previously been described elsewhere [1]. The background for the development of this system was also covered in this paper. A review of the current state of the art in characterising and classifying wheat quality by various means has been published [2]. A number of workers have previously attempted to classify wheat quality by odour [3-7].

Grain odour measurement is an indefinite science open to much interpretation. Whilst numerous standards are in place these all fail to provide surrogate or comparative standards against which an inspector may be trained or a sample compared. This is an undesirable situation since no reliable system is available when comparing the true provenance of samples in transport or in disputes over the quality of grain samples between supplier and end user. Hence, from a commercial standpoint the desirability of a reliable quality measurement system is clear. In addition to this a more important facet is the potential health implication that continually sniffing grain has on an inspectors health. It is becoming increasingly clear that there are significant risks in repeatedly sniffing samples contaminated by fungi (mycotoxin producing or not) and very small dust and grain particulates. It is against this background that the GASP system has been developed.

2 GASP construction

The system is based around an Osmetech 32 conducting polymer sensor array housed in a constant environment module (CEM, Osmetech plc) for better environmental stability. A dedicated sampling system has been built to carry odour from freshly sampled grain sealed in large (circa 30 g) disposable containers to the array with the minimum of obstruction (See Figs. 1 and 2).

Computer controlled valves are sequenced by the data acquisition software to allow variable sample exposure and backflushing of the system with a high humidity flow to wash the sensors and associated transfer lines to minimise carry over and contamination (see Figs. 1 and 2). The system is entirely enclosed, requiring the operator to simply present the sample to a holder and press the start button. An automated lift then presents a heated conditioning block to the sample and holds it for a defined period of time. The sample is then raised onto the needle assembly for headspace sampling and subsequent data acquisition.

The data acquisition routine incorporates a RBFann such that the software produces an on-line evaluation of the sample, producing a real-time output. The RBFann may be operated as a quantitative or classifying system. In quantitative mode it is able to interpolate between known classifications with the advantage that unknown samples that fall between example clusters are not forced towards one cluster or the other. The system was evaluated against commercial grain samples from 1998 and 1999 harvests.

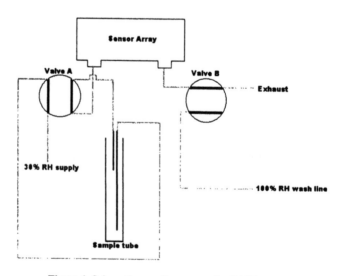

Figure 1 Schematic sampling system for GASP

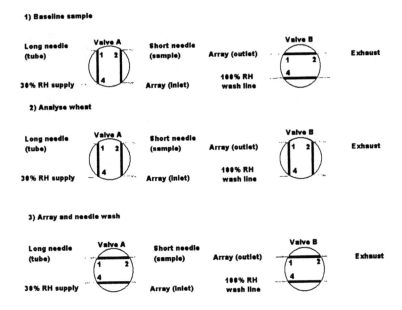

Figure 2 Position of valves during baseline, sampling and array wash operations

3. Performance

An odour panel trained at Silsoe Research Institute evaluated 1998 commercial grain samples. The samples were presented three times in random order and graded into very good, good, bad and very bad classes, the panel were asked to mark the samples 1-4 respectively. Samples of one variety from the 1998 harvest were used as the training set for the RBFann and a selection of 13 varieties, of wheat from the 1999 harvest were presented as the unknowns.

The odour panel classification of the 1998 grain is shown in Figure 3 with classes divided as good and bad, a grade mark of 2.5 being the threshold. The neural network, trained with these grain samples and classes was then used on-line with the electronic nose to classify the 1999 harvest grain.

The discrimination between good and bad grain types is not merely a result of differences in moisture content of samples. This is illustrated in Figure 4, where good and bad quality grains are almost equally distributed across the range of moisture contents encountered. Of the 50 samples analysed, the system correctly classified 38, the remainder were not graded by the system as bad or good. In the majority of cases this corresponded with an intermediate rating of the grain i.e. somewhere between good and bad.

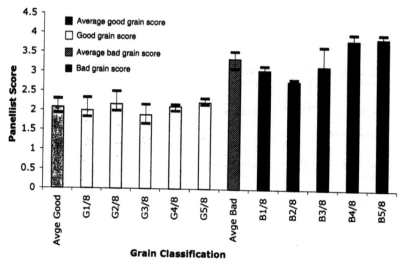

Figure 3 Odour panel classification of grain used in training the RBFann. The 'error bars' in the figure reflect the range of classification of each grain by the panel.

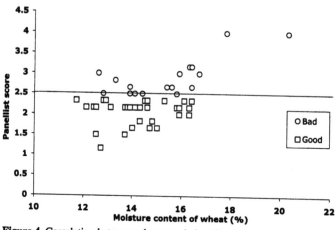

Figure 4 Correlation between odour panel classification and moisture content of the wheat

A PCA plot of the data, Figure 5, shows that the good and bad grain fall into distinct groups although the data could be considered as part of one elliptical cluster with subcategories within the cluster with opposite ends representing the best and worst grain samples.. The training data for the RBFann was collected with the GASP three weeks prior to classifying the unknown samples, indicating that sensor drift had minimal affect on the result. The PCA plot is merely a representation of the data for visualisation purposes, to give an indication of the 'sense' of the data.

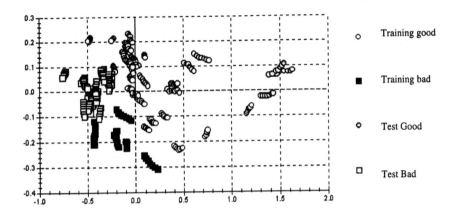

Figure 5 Principal component analysis (PCA) map of training and sample data (note the data consists of data slices rather than averaged data over the data acquisition period analysed).

   In this instance the simplistic view of grain as being either good or bad somewhat limits the data and forces the decision making into arbitrary choices. The benefit of the RBFann is that it produces an intermediate result. Further work with an enlarged data set should enable the grain to be re-evaluated against a more robust classification system such as good, intermediate good, intermediate bad and finally bad since the grain (as can be seen from the PCA plot) does not instantaneously transform from good to bad but follows a gradual transition from an 'optimal' good state through an intermediate stage and on to bad. However, the initial panel evaluated data set was not large enough to produce a reliable enough training set to produce sub-classified data against which real grain samples could be evaluated.

4. Conclusions

As can be seen from the results the GASP system is capable of classifying grain at a level equivalent to a trained odour panel. The classifications are independent of grain moisture content. Further development will investigate the robustness of the system over long periods of use and across a range of different grain samples. Grain quality classification into more groups would be a welcome improvement. The principal drawbacks of enlarging the number of grain classifications are the enlarged training set required, obtaining reliably classified training examples and the time involved in acquiring the data before real samples can be run. Development of the system to enable characterisation of other grain contaminants of interest e.g. pests such as weevils are planned.

   A dedicated system capable of determining the quality of wheat at points of transfer or of disputed quality has been developed. The dedicated nature of the system has enabled a more robust and user-friendly system to be developed. The design of the instrument has ensured that large samples can be repeatably classified. Important factors in the design are temperature and humidity control, consistent presentation of the sample, sensor cleaning and a quickly trainable neural network that is robust.

5. Acknowledgements

The authors wish to acknowledge the Ministry of Agriculture Fisheries and Food for funding this project (Project number MAFF CTD 9701) and the contribution of Osmetech plc in terms of loans of equipment and supply of expertise.

6. References

[1] Evans P,. Persaud K C, McNeish A S, Sneath R W, Hobson R N and Magan N Sensors and Actuators B, in press 2000

[2] Magan N and Evans P 2000 Journal of Stored Products Research 36(4) 319-340

[3] Stetter J R, Findlay Jr. M W, Schroeder K M, Yue C and Penrose W R 1993 Analytica Chimica Acta 284 1-11

[4] Pisanelli A M, Qutob A A, Travers P, Szyszko S, and Persaud K C 1994 Life Chemistry Reports 11 (1994) 303-308

[5] Jonsson A, Winquist F, Schnürer J, Sundgren H and Lundström I, 1997 International Journal of Food Microbiology 35 187-193

[6] Borjessön T, Eklöv A, Jonsson A, Sundgren H and Schnürer J 1996 Cereal Chemistry 73 457-461

[7] Borjessön T and Olsson J 1998 Proceedings of Symposium on Electronic Noses in the Food Industry, Stockholm, Sweden, 24-28

Section 5: Applications: Food, Agricultural and Environmental
Paper presented at the Seventh International Symposium on Olfaction and Electronic Noses, July 2000

217

Rancidity Investigation on Olive Oil: a Comparison of Multiple Headspace Analysis using an Electronic Nose and GC/MS

S. H. Hahn, M. Frank, U. Weimar

Institute of Physical and Theoretical Chemistry, University of Tübingen, Germany,

Tel: +49 7071 29 78765, Fax: +49 7071 29 5960

E-mail: simone.hahn@ipc.uni-tuebingen.de.

Abstract. The demand for fast and objective food quality control has led to extensive research in the field of sensor development. In this investigation fresh olive oil was aged artificially and the samples were measured with GC/MS and an Electronic Nose (MOSES), using a Multiple Headspace Extraction method (MHE). Based on the MHE the compounds generated during the oxidation process could be quantified. The results of the GC/MS measurements were used to track the volatile compounds and explain the Electronic Nose data, which were evaluated by PCA and quantified by PCR.

1 Introduction

Over recent years the demand for olive oil has increased, as olive oil is known to contain unsaturated fatty acids and antioxidants, which lower cholesterol and reduce the risk of heart attack. As a result of its diverse properties, olive oil is used in many countries where it is not necessarily traditional. A lot of Electronic Nose measurements have been carried out to discriminate between olive oils from different countries, species, branches, and processing (extra virgin, virgin) [1]. However, these investigations revealed a lack of reproducibility, which in most cases could be attributed to the natural spread of the samples and not to the measurement procedure itself. Many compounds detected by the Electronic Nose are generated by fermentation and storage; they are not necessarily specific to the olive plant or region. Consequently, the work focussed on only one type of olive oil, aged artificially with a UV lamp and measured with GC/MS and the MOSES system to monitor the volatiles generated by the oxidative decomposition of unsaturated fatty acids.

2 Experimental

For the Electronic Nose analyses the samples were investigated in a measurement set-up which comprises the hybrid Modular Sensor System (MOSES) [2] connected to a headspace sampler (HP 7694). The sensor array consists of eight tin oxide conductivity devices and eight polymer-coated quartz crystal microbalance oscillators. The GC/MS-headspace analyses were carried out by a gas chromatograph (HP 6890, column: HP-VOC) coupled with a mass selective detector (HP 5973 MSD) and a headspace sampler (HP 7694).

Samples were prepared by placing 2l of olive oil in a bottle and stirring them. To speed up degradation a UV-lamp (12W, $\lambda = 254$ nm) was immersed in the bottle for half an hour a day. Samples were taken after every two hours of UV exposure. Three samples were used for GC/MS measurements and eight for Electronic Nose measurements. The samples were equilibrated for 20 min at 90°C in the HS oven. A six step Multiple Headspace Extraction (MHE) was performed for every sample for GC/MS and MOSES.

For quantification and an exact qualification of the expected aldehydes with the GC/MS system, 13 different aldehydes (C_5 - C_{12}) which are known for their rancid smell [3] were submersed in olive oil and corn oil. The concentrations of the aldehydes were 100 ppm for aldehydes with higher vapour pressures (C_5 - C_6) and 1000 ppm for the ones with lower vapour pressures (C_8 - C_{12}). With these aldehydes in the oil matrices a six step MHE was carried out with the GC/MS and the MOSES II system.

3 Results

The results of the MHE-GC/MS measurements for the 13 aldehydes in oil matrices showed that the distribution coefficient* is not constant for the chosen parameters. For the first extraction step the distribution coefficient of nonanal and all other compounds is smaller than the one for the second extraction step. The reason could be that 20 min of equilibration is not enough and that linearity will be seen at a larger number of extraction steps.

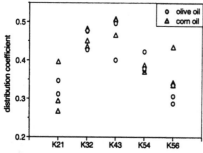

Figure 1: Distribution coefficient of 1000 ppm nonanal submersed in an olive and corn oil matrix (every 3 samples per matrix, 6 extraction steps).

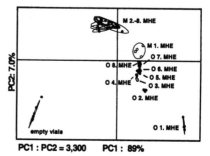

Figure 2: PCA of MOSES II measurements, performed with 13 different aldehydes submersed in an olive and corn oil matrix (8 extraction steps)

Figure 2 shows the result of a Principal Component Analysis (PCA) of a six step MHE of olive oil and corn oil samples, in which 13 different aldehydes were submersed. There is clear discrimination between the different matrices, olive oil and corn oil. The sensor signals for the first extraction step are quite different from the following ones. Looking at the signals, the values decrease by almost 50% from the first to the second extraction step. We suggested that water could be responsible for that effect and carried out some Single Ion Monitor (SIM) experiments with the GC/MS, where only water was monitored. For further extraction steps there is discrimination by the PC2 for olive oil samples. The pattern for corn oil is the same for all samples and no discrimination can be seen for the different extraction steps. The big difference between corn and olive oil for more extraction steps can be attributed to the different matrices. The matrix for olive oil is much more complex than the one for corn oil, which contains nearly no volatile compounds because of the manufacturing process. Because of these results of

---

* The distribution coefficient is defined as: K_{xy} = peak area (x) / peak area (y), where x and y are different extraction steps (x > y).

preliminary measurements we were optimistic that the MHE method would improve our results for aged olive oil.

Aging of olive oil: GC/MS measurements

The GC/MS measurements showed that mainly pentane and aldehydes ($C_5 - C_9$) originated from the oxidative decomposition of unsaturated fatty acids (Figure 3) [4].

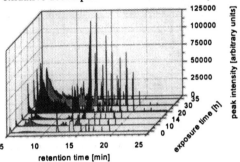

Figure 3: Chromatograms of olive oil samples of different UV exposure times.

The volatile compounds increasing during UV exposure are: pentane (6,90 min), pentanal (12.11 min), hexanal (14.10 min), 2-(E)-hexenal (15.09 min), heptanal (15.73 min), octanal (17.11 min), nonanal (18.35 min) and some other aldehydes in minor concentrations. Figure 4 shows the differences between the 1st and 6th extraction step of the MHE. There are differences in the peak areas for some components, such as pentane (6.4 and 6.9 min), pentanal (12.1 min), octane (13.9 min) and hexanal (14.14 min), which all are increasing.

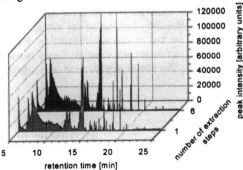

Figure 4: Chromatograms of the 1st and 6th extraction steps of a MHE for 35 hours aged olive oil samples.

Using the MHE [5] method and as a reference the results of the preliminary measurements (13 aldehydes submersed in olive oil) for GC/MS, it was possible to quantify the volatiles generated during the UV exposure (Table 1). The first aldehydes to be detected were hexanal and nonanal after 4 hours of exposure time. With the adjusted signal to noise ratio (s/n = 14) of the evaluation software package. Quantification was finally possible after 8 hours. The concentration of hexanal increases during the aging process more than the others. The high concentrations of hexanal, nonanal and 2-(E)-hexenal can be attributed to the high content of linolic (hexanal, nonanal) and linolenic acid (2-(E)-hexenal) in olive oil, which were quickly degraded. As by-product of the composition hexenylacetats are generated, which were also be found in aged olive oil (17,05 min).

exposure time [h]	pentanal [ppm] 12.12 min	hexanal [ppm] 14.14 min	2-(E)-hexenal [ppm] 15.09 min	heptanal [ppm] 15.72 min	2-(E)-heptenal [ppm] 16.59 min	octanal [ppm] 17.11 min	nonanal [ppm] 18.35 min
8		35 ± 0.8					8 ± 0.2
10		39 ± 1.7					9 ± 0.7
12		30.0 ± 0.8					2 ± 0.0
14		39 ± 0.9					7 ± 0.2
20	89 ± 17.7	429 ± 4.1	122 ± 18.2	59 ± 7.9	47 ± 1.0	11 ± 0.3	79 ± 1.5
25	132 ± 27.0	289 ± 0.8	60 ± 8.4	37 ± 5.1	6 ± 0.1	8 ± 0.2	51 ± 0.9
30	106 ± 22.0	407 ± 12.6	93 ± 13.6	82 ± 12.0	58 ± 1.2	14 ± 0.3	84 ± 1.6
35	53 ± 10.1	177 ± 4.6		28 ± 3.7		7 ± 0.7	33 ± 0.8

Table 1: Concentrations [6] of different aldehydes generated during the aging of olive oil.

The large deviations in the mean of some values in the concentration trend of hexanal and other components are due to the problem that some peaks were not integrated by the adjusted signal to noise ratio.

Calibration

The aging of olive oil by a UV lamp (30 min a day) took some time, so a calibration method for the sensors would have been necessary. As it is well known that changes in sampling can have a greater effect than errors in measuring, we were looking for highly reproducible calibration samples. The idea was to use a pressure bottle with a target analyte - for olive oil this would have been hexanal - and use it as a reference. That is possible because MOSES II is equipped with an Input-module, where a gas pump can suck a defined volume. Unfortunately hexanal is not stable under high pressures in a synthetic air atmosphere, so we had to choose propanal as reference gas (100ppm in synthetic air). As a second reference we used 100 ppm hexanal in olive oil and measured it in headspace vials. The sensor signals were very reproducible over the time.

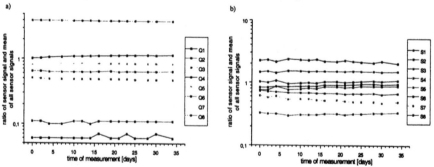

Figure 5: Averaged sensor signal related to the mean of all sensors of one Module (QMB (a) and MOX (b), sensor for 100 ppm propanal in synthetic air).

Deviations can be more attributed to fluctuations than to sensor drift [7] (mean standard deviations for the different sensor types: sdv (QMB) = 1.45%; sdv (MOX) = 4.07%).

Aging of olive oil, MOSES II measurements

The feature for evaluating the sensor signal was signal minus baseline for both sensor types. The result of the PCA of all samples (second extraction step) is shown in Figure 6. The MOSES II system is able to distinguish between samples aged a few hours (0-8h) and samples aged over 20 hours. For exposure times between 8 and 20 hours, no clear discrimination is seen in the PCA. This result is quite different to the results obtained from aging corn oil [8], where the discrimination between the different samples is quite good. The PCAs of different extraction steps looking nearly the same, the best discrimination being obtained with the second extraction step. The large deviation in the PC2 direction could be due to the error in sample taking. As shown above, the sensors are delivering reproducible data.

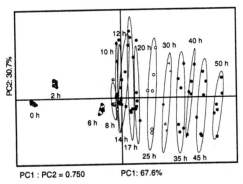

Figure 6: PCA of olive oil with different UV exposure times (2nd extraction step).

The GC/MS measurements show only slight changes, except for the jump between 14 and 17 hours, where the rancidity can also be detected by the human nose. The MHE provided a larger data set, which served as input for PCAs and PCRs. No significant differences in the results were seen in the PCA, but they were present in the PCR.

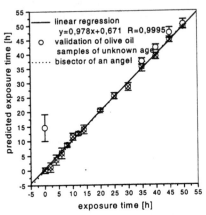

Figure 7: PCR of 1st MHE of aged olive oil samples.

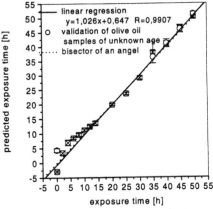

Figure 8: PCR of 2nd MHE of aged olive oil samples.

The PCR was carried out with 5 samples of each age using the cross validation method. Three samples per "unknown" age were predicted by the system. Figure 7 and 8 show the results of PCRs of the first and second extraction steps. For both calculations the correlation between predicted vs. measured is very good. The prediction of samples of "unknown" age for samples illuminated for a long time is better than that for those illuminated for a short time. For the first MHE the fresh olive oil was classified as 14 hour aged, which is definitely too high. The performance of a second extraction improves the results for fresh oil (4 hours). This improvement could be related to the lower water concentration in the headspace. As shown in the preliminary measurements, water has an influence (around 50%) on the sensor signal of the first extraction step. After the first extraction step the total water concentration decreases and the concentration of the other volatiles increases (Figure 4).

4 Discussion

The investigations of GC/MS and MOSES measurements showed that the increasing sensor signals are a result of increasing oxidation products such as pentane and hexanal. Other volatiles, which can normally be found in olive oil, such as ethanol, seem to play no role for discriminating between different aged samples. The MOSES system was able to classify aged samples according to low concentrations of off-odours. The MHE method offers some advantages by quantifying the aged olive oil samples. The explanation for this is that olive oil contains a small amount of water; at equilibration temperatures of 90°C, most of the water is in the headspace, thus the signals for the first extraction steps are influenced by water. However in the second extraction step there is a decreasing concentration of water in the headspace, resulting in sensor signals, which are up to 40% smaller (GC/MS, single ion monitoring).

References

[1]   M. Prado, G. Sberveglieri, S. Gardini and E. Dalcanale, Sequential classification of 14 olive oils types by an Electronic Nose, Proceedings of ISOEN 99, Tübingen 1999, 255-258.

[2]   J. Mitrovics, H. Ulmer, U. Weimer and W. Göpel, Modular sensor systems for gas sensing and odor monitoring: The MOSES concept, Accounts of Chemical Research, Vol. 31. No. 5 (1998), 307-315.

[3]   M. T. Morales, R. Aparicio and J. J. Rios, Dynamic headspace gas chromatograhic method for determine volatiles in virgin olive oil, Journal of Chromatography A, 668 (1994) 455-462.

[4]   M. T. Morales, J. J. Rios, R. Aparicio, Changing volatile composition of virgin olive oil during oxidation: flavors and off-flavors, J. Agric. Food Chem., 41 (1997) 2665-2673.

[5]   B. Kolb, L. S. Ettre, Static headspace-GC, theory and practice, Wiley-VCH 1997.

[6]   S. H. Hahn, Methoden zur Evaluierung der Qualität von Speiseölproben: instrumentelle Analytik im Vergleich mit Sensorsystemen, Diplomarbeit, Universität Tübingen 1999.

[7]   M. Frank, S. Hahn and U. Weimar, accepted for ISOEN 2000.

[8]   M. Frank, Vergleichende Messungen and Lebensmittelproben mit Verfahren der Analytischen Chemie und der Gassensorik, Diplomarbeit, Universität Tübingen 1998.

Section 5: Applications: Food, Agricultural and Environmental
Paper presented at the Seventh International Symposium on Olfaction and Electronic Noses, July 2000

223

Discrimination between Commercial and Local Olive Oils by Means of an Electronic Nose

S. Capone[§], A. Taurino[§], P. Siciliano[¥], R. Rella[¥], C. Distante[§],
L. Vasanelli[¥§], M. Epifani*

[¥]Istituto per lo Studio di Nuovi Materiali per l'Elettronica IME-CNR, Via Arnesano, 73100, Lecce, Italy Phone: +39 0832 320244, Fax: +39 0832 325299,

*INFM Via Arnesano, 73100, Lecce, Italy

[§]Dip. Ingegneria dell'Innovazione, University of Lecce, Via Arnesano, 73100, Lecce - Italy

Abstract. This paper presents the results obtained from the analysis of a variety of olive oils performed by means of an Electronic Nose based on an array of sol-gel tin oxide sensors. In particular good classification between commercial and not commercial local products has been obtained.

1. Introduction

One of the problems concerning food analysis is the characterization of olive oil according to the EU normative and its classification based also on the origin areas. Indeed for these areas it is important to receive the D.O.P. (Protect Origin Denomination) label which is the index of the high quality of the product. This is important for Apulia region, which is one of the main olive oil producers in Italy.

A considerable interest in this context is having the use of gas sensors array in combination with pattern recognition methods in order to identifing odours or gases since their poor selectivity. Selectivity can be improved by using an array of non-specific sensors with partially overlapping selectivity and a signal processing which may give a sensitive discrimination of the odours. It is proved that "Electronic Nose,, can be used as powerful method to complement the traditional chemical and sensorial analyses. Other methods like GC/MS allow obtaining highly accurate chemical analysis but are time consuming and uncorrelated with sensorial impact. In the sensorial analysis, a group of experts trained for this specific task and called panel test, provides information of the overall organoleptic quality of alimentary products. However, the sensorial analysis suffers from problems related to human judgement's subjectivity, which often depends on many physical and psychological factors [1].

The electronic nose instrumentation has been used to measure both simple and complex odours in many fields of application. Many researchers have reported on the use of an electronic nose to measure simple odours that arise from a single chemical compound. Among the fields, which can receive more benefits from this class of instruments, food analysis is the most important to satisfy the market demand. It is known that a fast and deep food inspection may be important for a better quality of life and production process quality control. The electronic nose used in food industry is only one example of complex odour discrimination analysis and has been widely reported and well approved. Some applications include the identification of different grades of wheat, types of cheese, grades of meat and the flavours of many kinds of drinks and beverages. [2]

For this reasons "Electronic Nose" must be trained with examples of odours to be recognised, so that the pattern recognition agent learns how to distinguish different aromatic patterns.

The Institute IME-CNR developed an electronic nose to test a number of different commercial olive oils of various qualities produced in different Italian areas. Furthermore, the analysed olive oils were compared with local non-commercial olive oils of the Salento area in Apulia.

Principal Component Analysis (PCA) has been used as a first analysis of the array gas sensor responses.

2. Experimental

In this work we used pure and Ni, Pd, Pt, Os-doped SnO_2 based thin film sensors prepared by sol-gel technology. Pure SnO_2 sols have been prepared starting from anhydrous $SnCl_4$ as precursor. For the doping solution preparation the prescribed amount of $NiCl_2 \cdot 6H_2O$, $OsCl_3$, $PtCl_2$ or $Pd(OOCCH_3)_2$ has been dissolved in suitable solvents and added to SnO_2 sol. The different sols have been deposited by spin coating from sol-gel solutions (5wt% for all dopants) at 3000rpm. The sensing layer has been deposited onto an allumina substrates (35×35) mm^2 pre-arranged to be cutted by obtaining (3×3) mm^2 single samples which are equipped at the back face by a Pt meander as heater. The films have been heated in air at various temperatures. All this process has been made twice in order to deposit 2 layers of sensing material. The film tickness was about 200 nm. Ti/Au interdigitated fingers were evaporated on the front face as electrical contact [3]. Sensors have been mounted onto TO-8 sockets and positioned in a test chamber and exposed to the olive oils volatile compounds. The sensor response towards volatile component of the aroma was measured by applying a constant dc voltage of 2 V to the electrodes and the electrical current measured by an electrometer (Keithley 6517A), equipped with a multiplexer to acquire the responses from all sensors in the test cell. A PC via Labview software controlled all the processes.

We used an experimental set-up based on static headspace measurements. Samples of different olive oils were introduced into a vial kept at room temperature by a thermostatic tank. During this phase, it is important that in the organic volatile compounds of samples the liquid and the gas phases are well balanced each other to reach a stable headspace. A mass flow control system (MKS mod 647B) was used to adjust the different mixtures composition trough mass flow controllers. The baseline has been acquired by using a mixture of dry air (flow 100 sccm) and nitrogen (flow 100 sccm) as a reference gas and performed the sensors conditioning. In this way the flavours were transferred collecting them by means of a deviation of the only nitrogen line as carrier gas into the array chamber but keeping the total flow at 200 sccm. For every one of the samples we made many measurements by fixing the exposure time at 15 minutes and a purging time of 20 minutes in synthetic air-nitrogen mixture. We chose the timing protocol after some preliminary studies about the baseline stabilisation, the array exposure time and the oil headspace stabilisation. It has been observed that longer time exposure does not increase the discrimination capability of the substances and this time is sufficient to obtain the saturation of the signal.

3. Results

We used two different groups of olive oils as test samples. The first group was composed of five commercial oils: 3 extra-virgin, one virgin and a husk olive oils.

These oils were randomly selected and were used for a first classification test. The second group was composed of five local non-commercial olive oils and was used for classification in relation to the place of origin.

The test chamber contains the different sensors in a symmetric shape around gas inlet. In the table 1 a scheme of the array composition is reported. With this sensors array we tested five commercial olive oil brand. Firstly, we tested five commercial olive oil brands that have been chosen in a randomly way. Their characteristics in terms of quality are reported in table 2.

Table 1 Sensor Parameter.

SENSORS	SnO_2	SnO_2/Os	SnO_2/Pt	SnO_2/Pd	SnO_2/Ni
SENSING MATERIALS	SnO_2	SnO_2	SnO_2	SnO_2	SnO_2
DOPANT	-	Os	Pt	Pd	Ni
THICKNESS (nm)	200	200	200	200	200

Table 2 Commercial oils used in the test.

OIL BRANDS	QUALITY
CARAPELLI	EXTRAVIRGIN
CORICELLI	EXTRAVIRGIN
DE SANTIS	EXTRAVIRGIN
DEL PAPA	VIRGIN
DENTAMARO	HUSK

As an example, figure 1 shows a typical plot of the dynamic response of the sensor array exposed to the olive oil headspace. It deals with the array current variation to five repeated exposures to the Coricelli oil. The response of the array for each experiment, in terms of percentage variation of current was calculate with:

$$S = \frac{I_{oil} - I_{baseline}}{I_{baseline}} \times 100$$

where I_{oil} and $I_{baseline}$ are current values at the end of the oil vapours exposure and in the reference gas.

PCA analysis was performed on raw matrix of responses to different oils through correlation matrix.

Figures 2 shows the first two components of the PCA plot. The plot shows the best results obtained at the sensor operating temperature of about 250°C. This analysis shows a good separation among clusters, not only for the different oil qualities (PCA plane is divided in tree different zones: extra-virgin, virgin and husk oil), but also for the classification of different oil brands. We can also note, in this figure, that over 99% of the total variance within the data is contained in the first two principal components.

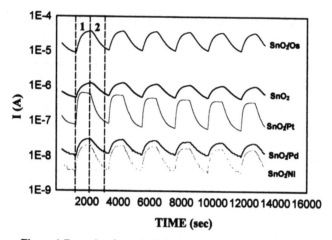

Figure 1 Example of a typical dynamic array response for an olive oil vapours. The zone labelled 1 represents the sensor responses of the aroma. The zone labelled 2 represents the recovery response in dry air – nitrogen atmosphere.

A second set of measurements analysed some local olive oils whose acidity degree was measured at the moment of their extraction. A classification of different non-commercial local olive oils related with their provenience areas and their acidity is described in table 3.

Figure 3 shows the PCA plot of the five local non-commercial olive oils. It is evident the good separation among clusters, and the electronic nose capability to distinguish oils from different areas. Furthermore, has been noted a preferential orientation in terms of the acidity degree.

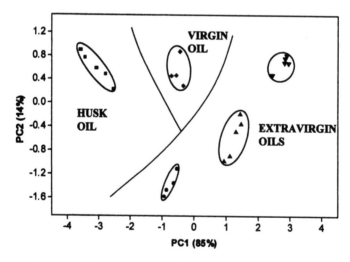

Figure 2 PCA plot of five different olive oil brands obtained by using sensors working at the temperature of 250°C.

Table 3 Local oils used in the oil volatile compound tests.

LOCAL OILS	PROVENIENCE AREAS	ACIDITY
OIL 1	Town: Scorrano Zone: Mustazzola	1
OIL 2	Town: Taviano Zone: 1	0.2
OIL 3	Town: Taviano Zone: 2	1.5
OIL 4	Town: Scorrano Zone: Suriani	2.2

Figure 4 shows the PCA plot of the overall analysed olive oils, where an interesting separation of the plane has been noted: one related to local oils and the other to the commercial one.

4. Conclusions

In conclusion sol-gel tin oxide sensors have shown good properties to be used for qualitative analysis of olive oils. In particular, an array of five differently doped tin oxide micro-sensors provides a good discrimination between commercial and local non-commercial olive oils produced in the Salento area. The sensors were also able to distinguish olive oils on tha basis of their quality: extravirgin, virgin and husk.

Future works will be addressed to the integration of electronic nose analysis with GC/MS and panel test.

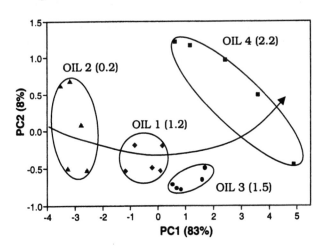

Figure 3 Principal component analysis of the sensors array for the five local olive oil.

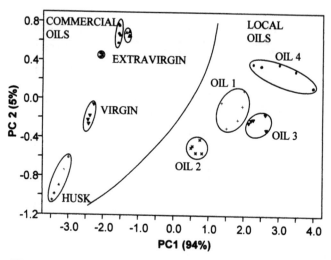

Figure 4 PCA plot of all the examined oils.

5. Acknowledgement

The authors would like to be greatful to Mr. Flavio Casino for his assistance during the measurements. This work was partially supported by the MADESS II finalized project of the CNR and FESR "Subprogram 2" EU project.

6. References

[1] J.W. Gardner, E.L.Hines and H.C.Tang. 1992 Sens. and Act. B 9 9-15.
[2] J.W. Gardner and P.N. Bartlett. 1998 Electronic Noses. Principal and Applications. Oxford University.
[3] F. Quaranta, R: Rella, P. Siciliano, S. Capone, M. Epifani, L.Vasanelli. Analysis of vapours and foods by means of an Electronic Nose Based on sol-gel metal oxide sensors array, Sens. and Act. B, to be published.

Section 5: Applications: Food, Agricultural and Environmental
Paper presented at the Seventh International Symposium on Olfaction and Electronic Noses, July 2000

229

Classification of Fresh Edible Oils with Piezoelectric Quartz Crystal Based Electronic Nose

Z.Ali, D.James, W.T.O'Hare, F.J.Rowell*, T.Sarkodie-Gyan, S.M.Scott, B.J.Theaker*

School of Science and Technology, University of Teesside, Middlesbrough, TS1 3BA
Phone: +44 (0) 1642 342463 Fax: 44 (0) 1642 342401
E-mail: z.ali@tees.ac.uk

*School of Sciences, University of Sunderland, Sunderland,
Tyne and Wear, SR2 7EE

ABSTRACT

An electronic nose based on an array of six coated Piezoelectric Quartz crystal (PZQ) sensors has been used for classification of fresh edible oils. The electronic nose was presented with 346 samples of fresh edible oil headspace volatiles, generated at 45°C. Extra virgin olive (EVO), olive oil (OI) and Sunflower oil (SFO), were used over a period of 30 days. The sensor responses were analysed using Principal Component Analysis (PCA), Simplified Fuzzy Adaptive Resonance Theory Mapping (SFAM), and Fuzzy logic classification (FLC). Results for SFAM and FLC were similar, both giving classifications above 95% for the test samples.

1. INTRODUCTION

The quality control of odours and volatile compounds is important in a number of sectors including food, chemical and the environmental [1]. With respect to fresh edible oils, the investigation of food volatiles is increasingly of interest. The most widely applied and established technique for their evaluation being Gas Chromatography coupled to Mass-Spectrometry (GC-MS) [2]. However, GC-MS is expensive and requires technical skill to operate.

There is a requirement for development of inexpensive portable instruments for online odour analysis at the factory floor. Electronic nose systems have been used for odour sensing [3], they may employ a variety of sensor types including Metal Oxide Sensors (MOS) [4], Metal Oxide Semi-Conductor Field Effect Transistor (MOSFET) [5], Conducting Polymer Sensors [6], Piezoelectric Quartz Crystal (PZQ) [7] and Surface Acoustic Wave Sensors (SAW) [8].

In PZQ sensors a chemical or biochemical layer allows extraction of analyte from a sample stream. The sensing is based on the change in device frequency (frequency shift), which is proportional to the mass of material sorbed by the coating on the crystal surface. The relationship between frequency shift and mass change was derived by Sauerbrey [9] and may be calculated using equation 1.

$$\Delta f = -2.3x10^6 . f_o^2 . \frac{\Delta M_s}{A}$$ (1)

Where Δf is the change in frequency of the quartz crystal in Hz.

f_o is the fundamental frequency of the quartz crystal in MHz.

ΔM_s is the mass of material deposited or sorbed onto the crystal in g.

A is the area coated in cm^2.

Martin *et al.*[4] have used a FOX 2000 instrument consisting of six MOS with Linear Discriminate Analysis to distinguish between different vegetable oils. A sensor array of four conducting polymers has been employed by Stella *et al.* [6] to characterise three commercial brands of extra virgin olive oil. Data analysis involving Principal Component Analysis was used to differentiate between the brands. Ulmer *et al.* [10] also used PCA to discriminate between different types of olive oil, but utilising a hybrid sensor system of eight coated PZQ and eight Semiconductor sensors.

In the present work we report on the use of a six-sensor array of coated PZQ for the detection of different commercial vegetable oils. PCA was used for visualisation; Simplified Fuzzy Adaptive Resonance Theory Mapping (SFAM) and Fuzzy Logic Classification (FLC) were used for classification.

2. EXPERIMENTAL

2.1 APPARATUS

The PZQ based six sensor array was developed with fundamental frequencies of 10MHz (SES Piezo Ltd., Portsmouth, Hants. UK). Each PZQ was coated with a commonly utilised gas chromatography stationary phase, each containing a different functional group to allow limited selectivity. The coated PZQ crystals are housed in a 10ml PTFE sensor chamber. A reference PZQ, allows for drift compensation. A PC collected the frequency of each PZQ via a PC-30AT interface card. Data acquisition, SFAM and FLC object oriented software was developed in-house in C++.

Three different types of vegetable oil were used for this work, Extra Virgin Olive Oil (La Chinata, 10600 Plasencia, Spain), Olive Oil (Filippo Berio, 55049 Viareggio (Luca) Italy), and Sunflower oil (Pure Golden, Associated Oil Packers, Bootle, UK).

2.2 PZ QUARTZ CRYSTAL COATING

The PZQ coatings used were OV-1, Carbowax 20M, OV-17, Diethylene glycol succinate, Silar 10C and OV-210 (Phase Separations Ltd). These were chosen to give a wide range of functional groups and polarities. Dilute solutions (0.1% w/w) of each coating were prepared in a volatile solvent, either chloroform ($CHCl_3$) or an 80:20 v/v mixture toluene:methanol. The solutions were applied to both sides of the crystal by means of a fine brush. The frequencies of the PZQ were monitored during the coating procedure so that frequency shifts were similar for each coating. The sensors were conditioned by passing nitrogen over the sensor array for six hours.

2.3 SAMPLING

10ml of oil in a 125cm^3 Dreschel bottle was stored at 45°C for 30 minutes to allow for headspace generation. During sampling the Dreschel bottle was kept in a water bath at 45°C to allow dynamic headspace analysis. A four-way PTFE valve was used to switch between the reference and sample gas, flow rate for both was set at 17ml min^{-1}. Sampling was performed over a 3-minute cycle, 1 minute base line reading (reference) and 2 minutes response (sample). After each reading the sample chamber was refreshed with reference nitrogen for 5 minutes prior to the introduction of the next sample. A total of 346 samples were taken over a 1-month period, consisting of 112 Extra Virgin Olive oil (EVO), 126 Olive oil (OI) and 108 Sunflower oil (SFO) samples.

3. Data analysis techniques

Three different types of data analysis were compared when classifying the vegetable oils using the electronic nose data.

3.1 Principal Component Analysis

Principal Component Analysis (PCA) was initially used for visualisation of the data. It is a commonly used multivariate technique [11], which acts unsupervised. This technique provides an insight into how effective the pattern recognition system will be at classifying the data.

3.2 Pattern Recognition Systems

Two pattern recognition systems were used to classify the data, namely fuzzy logic (FLC) and Simplified Fuzzy ARTMAP (SFAM). The attributes of the analyte (frequency change for each sensor) were extracted and stored in a feature matrix. Both of the classification methods were tested with the same feature sets.

For a classification regime to be effective the training data must fully satisfy two criteria:
- Every class must be represented.
- For each class, statistical variation must be adequately represented.

In general a large number of training sets will allow for noise effects if these are truly random. If the noise is not random then the regime will learn the noise pattern, possibly masking the true data patterns. If the data classes are well separated then few training sets may be needed to adequately describe the pattern, however if there are classes that fall near a decision boundary then it is important to use a larger number of data sets from near that boundary.

3.2.1. Fuzzy Logic

Fuzzy logic based on fuzzy set theory was employed to discriminate between gaseous organic compounds (classification). The classifier used Radial based functions (Guassian) to describe the data in fuzzy terms. Each fuzzy set has a mean frequency and a standard deviation of that frequency. Each sensor used has such a fuzzy set for each compound under investigation as shown in **figure** 2.

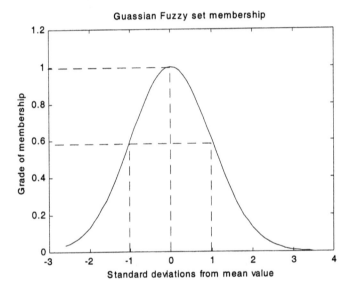

Figure 2 Guassian Fuzzy Set

The grade of membership μ (x) for a given frequency change from the base value is given by

$$\mu(x_i) = \exp\left[-\frac{(x_i - \bar{x}_i)^2}{2\sigma_i^2} \right] \quad \text{where } i = 1, 2...,n \tag{2}$$

where $\bar{x}$ is the mean of x and σ is the standard deviation.

For each compound to be classified a vector of fuzzy set values (one for each sensor) was constructed. The procedure is similar to that of Yea *et al.* [12]

$$CG = \begin{array}{c} f_1 \\ f_2 \\ f_3 \\ . \\ f_n \end{array} \begin{bmatrix} \mu_{GO} & \mu_{GE} & \mu_{GS} \\ \mu_{O1} & \mu_{E1} & \mu_{S1} \\ \mu_{O2} & \mu_{E2} & \mu_{S2} \\ \mu_{O3} & \mu_{E3} & \mu_{S3} \\ . & . & . \\ \mu_{On} & \mu_{En} & \mu_{Sn} \end{bmatrix} \quad (3)$$

where μ_{GO} is the fuzzy group set for Olive oil, μ_{GE} for Extra Virgin Olive oil and μ_{GS} for Sunflower oil.

then $CG = \{\mu_{GO}, \mu_{GE}, \mu_{GS}\}$ \hfill (4)

The defuzzification method used was that of the mean of height [13][14].

$$\mu_{Gx} = \frac{1}{n} \sum_{i=1}^{i=n} \mu_{Gxi} \quad (5)$$

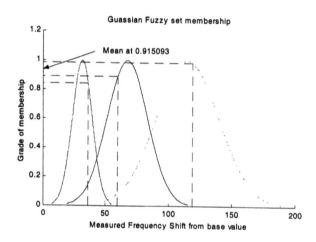

Figure 3 Mean of Height Defuzzification

Equation 5 and figure 3 show the mean of height defuzzification method for an array of n sensors. The classification category = CG α, where α = maximum of the fuzzy set CG.

$$CG\ \alpha = \mu_{GO} \cup \mu_{GE} \cup \mu_{GS} \quad (6)$$

This method also gives a level of agreement for each compound, a perfect match being 1.

3.2.2. Simplified Fuzzy ARTMAP (SFAM)

Simplified Fuzzy Adaptive Resonance Theory Mapping is a specialised form of a neural network for pattern classification [14][15]. The network has a small number of parameters and does not require guesswork to determine the initial configuration since the network is self-organising. In a standard back-propagation network used for pattern classification an output node is assigned to every class of object that the network is expected to learn. In SFAM the assignment of output nodes to categories is left to the network. SFAM carries out supervised learning in a similar way to a back-propagation network, but is more sensitive to noisy data. If the network vigilance parameter is initially set too high the network is less able to cluster data, mapping an output node to each input vector, becoming a look-up table [16]. The network is however self-organising, self-stabilising and suitable for real time learning [17].

4. RESULTS AND DISCUSSION

4.1.1. Principal Component Analysis

All of the readings were processed using (Minitab 12.1) for principal component analysis. The data shown in **figure 2** shows the clustering for the three types of oil. These data clusters indicate that the classification techniques would work well as the plot clearly shows three clusters with a small degree of overlap.

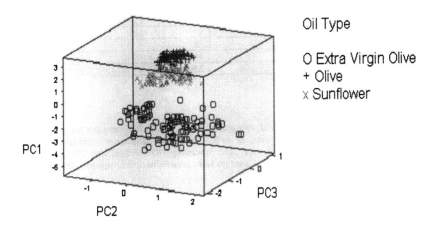

Figure 2 Principal Component Analysis of vegetable oil data

4.1.2. Simplified Fuzzy ART Mapping

The data was initially split into two groups, training and testing. 233 data sets were used for the training set and 113 for the test set. The results are shown in a confusion matrix (**table 1**). A high classification rate of 99% was achieved, there was one incorrect prediction; an extra virgin olive oil sample was classified as sunflower oil.

4.1.3 Fuzzy Logic Classifier

The FLC performed a high classification rate of 99% of the test set; one sunflower oil test sample was incorrectly classified as olive oil (**table 2**). Both the SFAM and the FLC produce comparable results for the classification of the vegetable oils. This was expected, as both are fuzzy techniques, and the data pre-processing similar. It may be seen that the misclassifications appear to occur around the boundary of the sunflower oil. With more data obtained from this area, it is reasonable to assume that the classifiers could perform higher classification rates.

Table 1. Confusion Matrix of oil data results using SFAM

Prediction	Actual			Prediction Total	Prediction Error %
	Extra-Virgin Olive	Olive	Sunflower		
Extra Virgin Olive	36	0	0	36	0.00
Olive	0	42	0	42	0.00
Sunflower	1	0	34	35	2.86
Actual Totals	37	42	34	113	0.88
Actual Error %	2.70	0.00	0.00	0.88	

Table 2. Confusion Matrix of oil data results using FLC

Prediction	Actual			Prediction Total	Prediction Error %
	Extra-Virgin Olive	Olive	Sunflower		
Extra Virgin Olive	37	0	0	37	0.00
Olive	0	42	1	43	0.00
Sunflower	0	0	33	33	3.03
Actual Totals	37	42	34	113	0.88
Actual Error %	0.00	0.00	2.94	0.88	

5. CONCLUSION

The classification of different types of vegetable oil using a coated PZQ array and fuzzy classification methods produced discrimination rates of 99%. Two new classification techniques have been successfully applied to electronic nose data processing. The SFAM and FLC have proven understandable with fast classification of complex food components such as vegetable oils.

6. ACKNOWLEDGEMENTS

This work was partially supported by HEFCE and ERDF under contract 70/41/001. The authors are grateful for the help in construction of the system by J. Legg and K. Middleton.

7. REFERENCES

[1]   Ali Z, O'Hare W.T, Sarkodie-Gyan T, Theaker B.J. and Watson E. 1999 Proc. SPIE, vol. 3853, 116-120.
[2]   Hiserodt R.D, Ho C.T, Rosen R.T. 1997 *ACS SYM SER*, 660, 80-97
[3]   Persaud K and Dodd G.H. 1982 *Nature*, 299, 352-355
[4]   Martin Y.G, Pavon J.L.P, Cordero, B.M. and Pinto C.G. 1999 *Anal.Chim.Acta*, 384, 83-94
[5]   Paulsson J.P. and Winquist F. 1999 *Foren.Sci.Int.*,105, 2 (1), 95-114
[6]   Stella R, Barisci J. N, Serra G, Wallace G. G, DeRossi D. 2000 *Sensors and Actuators B*, 63, 1-9,
[7]   Ali Z, O'Hare W.T, Sarkodie-Gyan T, Theaker B.J. 1999 *J. Therm Anal Calorim*, 55 (2) 371-381
[8]   Fang M, Vetelino K, Rothery M, Hines J, Frye G.C. 1999 *Sensors and Actuators B*, 56 (1-2) 155-157
[9]   Saubrey G.Z. 1959 *Z. Phys.*, 155, 206
[10]   Ulmer H, Mitrovics J, Noetzel G, Weimer U, Göpel U. 1997 *Sensors and Actuators B*, 43, 24-33
[11]   Byun H.G, Persaud K.C, Khaffaf S.M, Hobbs P.J, Misselbrook T.M. 1997 *Computers and Electronics in Agriculture*, 17, 233-247
[12]   Yea B, Konishi, R, Osaki, T, Sugahara K. *1994 Sensors and Actuators A 45 159-165*
[13]   Drainkov D, Hellendoorn H, Reinfrank M. 1996 An Introduction to Fuzzy Control 2nd Ed. Springer
[14]   Kosko B. 1992 Neural Networks and Fuzzy Systems, Prentice-Hall, Englewood Cliffs, NJ
[15]   Kasuba T. 1993 Simplified Fuzzy ARTMAP. *AI Expert, 8, November 18-25*
[16]   NeuNet Pro User Guide, Revision 2.2. 1999 CorMac Technologies Inc
[17]   Llobet E, Hines E.L, Gardner, J.W, Bartlett P.N, Mottram T.T. *1999 Sensors and Actuators B 61 183-190*

Section 5: Applications: Food, Agricultural and Environmental
Paper presented at the Seventh International Symposium on Olfaction and Electronic Noses, July 2000

235

Electronic Nose Based on Conducting Polymers for the Quality Control of the Olive Oil Aroma. Discrimination of Quality, Variety of Olive and Geographic Origin

A. Guadarrama[1], M.L. Rodríguez-Méndez[2], C. Sanz[3], J.L. Ríos[3], J.A. de Saja[1]

[1]Dpto. Física de la Materia Condensada. Facultad de Ciencias. Prado de la Magdalena s/n. 47011 Valladolid. Spain
[2]Dpto. de Química Inorgánica. E.T.S. Ingenieros Industriales. Universidad de Valladolid. Pº del Cauce s/n. 47011 Valladolid. Spain. e-mail:mluz@dali.eis.uva.es
[3] Instituto de la Grasa . CSIC. Avda. Padre García Tejero, 4, 41012 Sevilla.

Abstract. An electronic nose purposely designed for the organoleptic characterisation of olive oil is reported. The instrument works using an array of electrodeposited conducting polymer based sensors. Such an array has been able to distinguish not only among olive oils of different qualities (extra virgin, virgin, ordinary and lampante) but also among Spanish olive oils of different geographic origins and olive oils prepared from different varieties of olives.

1. Introduction

The detection of aroma volatiles emitted by olive oils is of key importance in the quality control of this product. Physico-chemical techniques (GC, GC-MS, HPLC), chemical analysis and sensorial analysis (panel test) are the classical methods used for this purpose, but they are expensive, time consuming and do not allow on-line measurements [1].

Many publications report the application of different prototypes or commercial devices for odour differentiation of food products including olive oil [2,3]. Nevertheless, it is clear now that systems specific for a particular application can be more successful than a single system suitable for all applications [4-6].

Our group is interested in developing an electronic nose specific for the discrimination of olive oils with different organoleptic characteristics. The commercially available instruments based on inorganic oxides have shown a limited utility in this particular case. For instance, in a recent work, a commercial instrument has been successfully used to distinguish among sunflower oil and olive oil, which is a trivial case [2]. In addition, most of these works have a lack of rigour in the selection and characterisation of the samples. This fact makes difficult to compare effectively the efficiency of the new method, the electronic device, with the classical chemical analysis or the panel test.

In previous works, sensor arrays formed by conducting polymers have shown to be able to discriminate among olive oils with well defined organoleptic characteristics [7,8].

In particular, our group developed a purposely designed sensor array formed by eight polymeric sensors, devoted to the evaluation of the olive oil aroma [7]. Such a prototype was able to discriminate among olive oils of different qualities (extra virgin, virgin, ordinary and lampante). In this work, the array has been extended up to sixteen sensors where new sensors with better sensitivity, reproducibility, repeatability and selectivity have been included.

The performance characteristics of the electronic device has been tested by using a selection of well defined samples (different varieties of olive, geographic origins, organoleptic characteristics, etc). Such a selection has been carried out by experts in the field, with the collaboration of olive oil producers. The results have been compared with those obtained by chemical analysis and with sensory analysis.

2. Experimental

2.1. Olive oil samples, sensory and chemical analysis

The olive oil samples under study include samples of two varieties of olive (Picual and Hojiblanca), samples from different geographic areas (from south, south-east of Spain and one from Portugal).

The olive oil samples were stored at −20°C under dark and unfrozen just before use.

The sensory analysis (carried out by a panel of experts) was performed following the rules of the International Olive Oil Council [9]. Te positive attributes and defects perceived in the virgin olive oil can be identified as fruity green, fruity ripe, apple, green leaves, bitter, pungent, winey, rancid, musty fusty and muddy. After the quantitative descriptive analysis, (QDA) and taken into account the overall impression, each panel judge gives an overall score from the oil.

The chemical analysis included the measurement of several parameters. The free acidity, which is indicative of the free fatty acid content of the oil expressed as oleic acid [10]. The peroxide value, which is a measure of the amount (meq O_2/Kg) of hydroperoxides formed through autooxidation during storage [11]. Finally, the absorbance at 232 and 270 nm provides a measurement of the state of oxidation of the oils [12].

2.2. Preparation of the polymeric sensors

Electropolymerisation and electrochemical measurements were performed using an EG&G PARC Model 263 potentiostat/galvanostat, which was controlled by electrochemical software M270/250 installed in a desktop computer. The reference electrodes were Ag/AgNO$_3$ 0.1 mol· L^{-1} (EG&G) for the non-aqueous media, and Ag/AgCl when water is used as solvent. All potentials quoted are relative to the corresponding reference. The counter electrode was a large surface area platinum gauze, which was flamed prior to use. The solutions were prepared and introduced into a cell with a thermostatic jacket (Metrohm) and with a temperature controlled liquid system (Neslab). All the polymeric films were grown at a constant temperature (25°C). The solutions were deoxygenated by bubbling nitrogen for 10 minutes prior to use.

The poly(3-methylthiophene) (PMT) and polyaniline (PAN) films were grown electrochemically onto glass substrates covered with ITO electrodes (electrode spacing 75 μm). The poly-pyrrole (PPy) sensors were deposited onto alumina substrates covered with gold electrodes (electrode spacing 50 μm).

Table 1. Conducting polymer sensors formed by electrochemical polymerisation using different techniques: chronoamperometry (CA) and cronopotentiometry (CP)

Sensor	Monomer	Electrolyte	Technique	Sensor	Monomer	Electrolyte	Technique
S01	Aniline	CF_3-COOH	CA	S09	Aniline	$HClO_4$	CA
S02	Aniline	CF_3-COOH	CA	S10	Aniline	$HClO_4$	CA
S03	Aniline	HBF_4	CA	S11	3-methylthiophene	$TBABF_4$	CP
S04	Aniline	HBF_4	CA	S12	3-methylthiophene	$TBAClO_4$	CP
S05	Aniline	HCl	CA	S13	3-methylthiophene	$LiCF_3SO_3$	CP
S06	Aniline	HCl	CA	S14	3-methylthiophene	$LiClO_4$	CP
S07	Aniline	HNO_3	CA	S15	Pyrrole	$K_4Fe(CN)_6$	CA
S08	Aniline	HNO_3	CA	S16	Pyrrole	$K_4Fe(CN)_6$	CA

The PAN sensors were generated from solutions of aniline (1 mol· L^{-1}, Aldrich) in deionised water (Milli-Qplus, Millipore), using CF_3COOH (Aldrich), HBF_4 (Fluka), HCl (Panreac), HNO_3 (Panreac) and $HClO_4$ (Fluka) as doping agents. In all cases the concentration of dopant anions was 2 mol· L^{-1} and the sensors were deposited by chronoamperometry (CA) technique. The films were generated at a constant potential of 0.9 V for 120s. Two different conditioning potentials (-0.6V and +0.6V) was finally applied to the polyaniline films, in order to get polymeric films in different oxidation states.

The PMT sensors were obtained from an electrolytic solution of 3-methylthiophene (0.1 mol· L^{-1}, purchased from Sigma), and with 0.1 mol· L^{-1} of the corresponding salt in acetonitrile (Sigma-Aldrich, HPLC grade). Lithium perchlorate anhydrous (Fluka), lithium trifluoromethane sulfonate (Fluka), tetrabutylammonium perchlorate (Sigma) and tetrabutylammonium tetrafluoroborate (Fluka) were employed in the polymerisation reaction to check the effects of dopant anions. These polymeric sensors were prepared by chronopotentiometry (CP) at a constant current (0.6 mA for 60 s). Under these conditions, stable oxidised blue films are obtained. In a subsequent conditioning stage, a –0.5 V potential was applied to the films in order to obtain the reduced state.

The PPy films were obtained from an aqueous solution of pyrrole (0.5 mol· L^{-1}, Aldrich) using potassium ferrocyanide (II) ($K_4Fe(CN)_6$· $3H_2O$) (0.1 mol· L^{-1}, Aldrich) as doping agent. These sensors were prepared by chronoamperometry , at a constant potential of 0.8 V for 60 seconds. Two different conditioning potentials (-0.6V and +0.0V) were applied to the polypyrrole films, in order to get polymeric films in different oxidation states.

The sixteen polymeric sensors used in this work, as well as the growth techniques and the preparation conditions, are collected in Table 1. Once prepared, the polymeric films were washed and stored in a desiccator chamber until use.

2.3. Electronic nose measurements

The analyses were conducted with a purposely designed system combining an array of 16 conducting polymer gas sensors with a static headspace autosampler unit (H.P. Mod. HP7694e). The complete device was monitored by a dedicated software including PCA.

A certain mass of olive oil (2.0 g) are placed in 10 mL vials for the static headspace sampling. The vials are encapsulated and placed in an HP7694e automatic headspace sampler from Hewlett-Packard. The vials are kept at a constant temperature (40 °C) for 9 minutes in order to obtain a homogeneous headspace. The vials are then pressurised for 8 seconds at 1.5 bar. The pressure gradient that builds up permits to fill a 3 mL loop in 9 seconds, and its content is then injected to the sensor chamber. The carrier gas was high purity synthetic air (99.9990%) and the flow rate was 100 mL· min⁻¹. An injection was performed every 15 minutes.

The sensors are mounted in a stainless steel text box with a volume of approximately 15 mL. A flow of dry air is set for 1 minute in order to stabilise the baseline. The carrier gas passes then through the sample loop for one minute, dragging the volatile components of the sample towards the chamber of sensors. Finally, a clean air flow is kept for 15 minutes in order to revert to the baseline.

As the polymeric gas sensors produce immediate responses, a fast data collection procedure is required. This step is performed through a PC-LPM-16 data acquisition card from National Instruments interfaced to a personal computer. The sensors are polarised using a constant voltage of 5V provided by a Hameg 8142 programmable power supply. The scan rate used to measure the changes of resistance of the sixteen sensors is 0.2s. The data are monitored in real time and the graphs can be followed using Visual Basic software from Microsoft.

The obtained data are exported to the GPES44 program, and the different features of the signals are analysed. The height, full width at half height and area of the signals, as well as the area of the derivative of the responses are routinely determined. Finally, pattern recognition techniques have been used for the discrimination of the signals. For this purpose Principal Component Analysis (PCA) has been carried out using the software Matlab v4.2.

3. Results and discusión

3.1. *Sensory and chemical analysis of the olive oils*

The panel of experts scored and described the olive oils under study. The results of the test are shown in Table 2. Regarding the panel of experts, it is important to remind that the highest the score obtained by an olive oil, the highest the quality. According to the scores given by the panel of experts, the set of samples also includes olive oils of different qualities: extra virgin (highest quality), ordinary (medium quality) and lampante (lower quality).

The results of the chemical analysis including the acidity, the peroxide index, and the absorbance parameters K232 and K270, have also been included in the Table 2. The acidity and peroxide index show a good correlation with the scores given by the panel of experts. As expected, the highest are the acidity and peroxide indexes; the lowest is the quality of the oil.

The olive oil samples were also characterized by gas chromatography. The headspace of the lampante olive oil is more complex than that of the extra virgin olive oil and the higher concentration of volatiles must be detected by the array.

Table 2. Samples of virgin olive oil from different origin, variety and category. Also a chemical characterisation of samples have been included.

Sample	Origin	Variety	Panel score	Category	Acidity	Peroxide value	K_{232}	K_{270}
I	Puente Génave (Jaén)	Picual	7.40±0.5	Extra virgin	0.20	8.30	1.80	0.12
II	Mengibar (Jaén)	Picual	7.19±0.5	Extra virgin	0.27	9.00	1.87	0.12
III	Antequera (Málaga)	Hojiblanca	5.95±0.5	Ordinary	0.76	8.06	1.70	0.11
IV	Puente Génave (Jaén)	Picual	5.00±0.5	Ordinary	0.20	9.30	1.78	0.12
V	Mengibar (Jaén)	Picual	4.72±0.5	Ordinary	1.23	39.20	2.19	0.16
VI	Valed´Alba (Castellón)	Blend	3.89±0.5	Ordinary	3.86	16.30	2.92	0.20
VII	lisboa (Portugal)	Blend	3.30±0.5	Lampante	1.28	10.50	2.15	0.19
VIII	Valed´Alba (Castellón)	Blend	2.45±0.5	Lampante	7.76	18.70	3.28	0.34

3.2. *Response of the polymeric sensors to the olive oils.*

Figure 1 shows the typical response of the array of sensors towards the olive oils headspace. The exposure to the olive oil aroma causes a fast, reversible and reproducible change of the conductivity of the polymeric sensors. In addition, the different sensing units show different sensitivity towards the olive oil headspace. For the poly(3-methyltiophene) sensors, increments in the resistances were observed, whereas the sensors produced from polyaniline and polypyrrole showed the opposed tendency. The baseline is completely recovered in all cases, although this process is very fast in poly(3-methyltiophene), slightly slower in polyaniline, and more time consuming for polypyrrole.

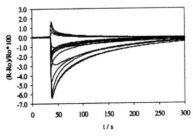

Figure 1. Response of the array of sensors to a lampante olive oil (sample number VII)

3.3. *Response of the array of sensors towards the olive oil samples*

The array of sensors was repeatedly exposed to the olive oils. Five measurements were carried out. Nevertheless, the first measurement differed from the subsequent and was discarded systematically. The peak height data were used to extract the Principal Component. In a first set of experiments, ordinary olive oils of similar characteristics but different geographic origin and variety of olive were exposed to the array (samples III, IV, V and VI). As shown in Figure 2, the 90% confidence ellipses are perfectly separated from each other. The largest part of the information is reduced to the first principal component (86.7%), which is also the most important in the discrimination of the clusters. The second principal component comprises a lower amount of information (10.8%). A certain tendency to change is observed for this parameter, that has previously been attributed by other researchers to the losses in the headspace of the vials after repeated measurements.

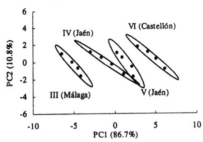

Figure 2. PCA of olive oils of medium quality (ordinary olive oils) with different geographic origin

The two lampante oils with different geographic origin were also analysed. Again, five measurements have been performed, and the PCA has always lead to a clear discrimination of the samples. The oil from Portugal has much higher values of the first principal component than that coming from Castellón. These results are depicted in Figure 3, where an ordinary olive oil from Málaga (sample III) has been included for comparison purposes.

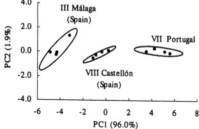

Figure 3. PCA of olive oils of low quality (lampante olive oils) with different geographic origin

4. Conclussions

The response of the array of polymeric sensors, associated with multivariate analysis methods, facilitates the clustering and differentiation of the examined olive oils.

A prototype of electronic nose using a sensor array based on conducting polymers purposely designed for the detection of olive oil aroma has been successfully used for the discrimination of very similar types of oils whose only difference was the geographic origin inside a small area.

5. Acknowledgements

Financial assistance from CICYT of Spain (Grant n° OLI-96-2172) is gratefully acknowledged. One of us (A.G.) would like to thank the Junta de Castilla y León for a grant.

6. References

[1]   Olias J M and Gutierrez F 1980 Grasas y Aceites 31 391-395
[2]   González Y, Pérez J L, Moreno B and García C 1999 Anal. Chim. Acta 384 83-94
[3]   Bartlett P N, Elliot J M, Gardner J W 1997 Food Technology 51 44-48
[4]   Guadarrama A, Fernández J A, Íñiguez M, Souto J, de Saja J A 2000 Anal. Chim. Acta 411 193-200.
[5]   Ulmer H, Mitrovics J, Noetzel G, Weimar U and Göpel W 1997 Sensor and Actuators B 43 24-33
[6]   D. Hodgins 1995 Sensors and Actuators B 26-27 255-258
[7]   Guadarrama A, Rodríguez Méndez M L, de Saja J A, Ríos J L and Olías J M Sensors and Actuators B (in press)
[8]   Stella R, Barisci J N, Serra G, Wallace G G and De-Rossi D 2000 Sensors and Actuators B 63 1-9
[9]   IOOC 1985 International Trade Standard applying to olive oils and olive-residue oils (COI/T.15/No.1), International Olive Oil Council (Athens)
[10]   IUPAC 1979 method N° 2,201, 7th edition
[11]   IUPAC 1979 method N° 2,501, 7th edition
[12]   CAC/RM 1970 method 26

Section 5: Applications: Food, Agricultural and Environmental
Paper presented at the Seventh International Symposium on Olfaction and Electronic Noses, July 2000

241

An Automatic Olfactory System for the Assessment of the Quality of Spanish Olive Oils

J. Martinez[1], G. Pioggia[2], F. Di Francesco[2], M.L. Rodríguez-Méndez[3], J.A. de Saja[1]

[1]Dpto. Física de la Materia Condensada. Facultad de Ciencias. Universidad de Valladolid. Prado de la Magdalena s/n. 47011 Valladolid. Spain. e-mail: jamar@fmc.uva.es
[2]Centro E. Piaggio. Facoltà di Ingegneria. Università di Pisa. Via Diotisalvi 2. 56126 Pisa. Italy
[3]Dpto. de Química Inorgánica. E.T.S. Ingenieros Industriales. Universidad de Valladolid. Pº del Cauce s/n. 47011 Valladolid. Spain. e-mail:mluz@dali.eis.uva.es

Abstract. A specific automatic system for the detection of organic volatile compounds responsible for the aroma of Spanish olive oil is presented. The system is based in a selected array of tin oxide sensors that present the best performance when are exposed to different olive oils. The device allows an accurate exposure of the sensors to the olive oil headspace, and collect the changes in the resistance of the sensors. The designed software controls the data acquisition and advanced statistical methods allow the classification of the Spanish olive oils.

1. Introduction

The food industry has the lack of slow and objective analytical methods for the monitoring of the organoleptic quality of their products. The present methodology has a high economic cost and is oriented to general purposes.

Nowadays the electronic nose technology, based on the use of sensor arrays, has demonstrated to be capable of distinguishing and identifying certain odours in a number of different applications [1-2]. Nevertheless, these systems have hitherto generally shown an inadequate behaviour when used in specific areas of food industry.

During several years our group has been working with a range of sensors that are characterised by changes in their resistance in presence of aroma [3-4]. The differences in the responses of the sensors have led us to develop an automatic olfactory system specifically designed for the monitoring of aromatic components in Spanish olive oil [5].

The system is based in a selected array of tin oxide sensors that present good performances when exposed to olive oil samples with different organoleptic characteristics. The device allows an accurate exposure of the sensors to the olive oil headspace, and collect the changes in the resistance of the sensors. An specific designed software controls the data acquisition and advanced statistical methods allow the classification of the olive oils.

The hardware of this system and the software designed for the control of the different units and for data processing are described.

2. Experimental

The automatic olfactory system has been designed to identify and classify the information extracted from the analysis of the volatile organic compounds (VOCs) of olive oils. The principal components of this system can be divided in different sections: sensors and samples, exposure system, measurement unit, and data acquisition. Figure 1 shows a general view of the automatic olfactory system

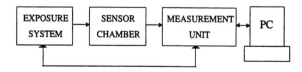

Figure 1 Automatic olfactory system

2.1. *Sensor and samples*

In a preliminary work, a series of tin oxide sensors were tested with different olive oil samples. The best performance was achieved with the array of sensors: TGS822, TGS826, TGS880, TGS2180 and TGS2620 from Figaro Inc. Up to 15 well organoleptically characterised olive oil samples were provided by the Institute of Fats of Seville (Spain) for the sensor selection. The different samples used are described in the Table 1.

Table 1 Spanish olive oil samples

Sample	Variety	Acidity	I.P. meq/Kg	K232	K270	P. Score	Quality	Origin
m01	Hojiblanca	0.17	7.05	1.89	0.14	8.15±0.5	Extra	Antequera (Málaga)
m02	Hojiblanca	0.76	8.06	1.70	0.11	5.95±0.5	Ordinary	Antequera (Málaga)
m03	Mezcla	3.86	16.30	2.92	0.20	3.89±0.5	Ordinary	Valed´Alba (Castellón)
m04	Mezcla	7.76	18.70	3.28	0.34	2.4±0.5	Lampante	Valed´Alba (Castellón)
m05	Picual	0.27	9.00	1.87	0.12	7.19±0.5	Extra	Mengibar (Jaen)
m06	Picual	1.23	39.20	2.19	0.16	4.72±0.5	Ordinary	Mengibar (Jaen)
m07	Picual	0.20	8.30	1.80	0.12	7.40±0.5	Extra	Puente Génave (Jaén)
m08	Picual	0.19	8.30	1.81	0.12	7.6±0.5	Extra	Puente Génave (Jaén)
m09	Picual	0.20	9.30	1.78	0.12	5.0±0.5	Ordinary	Puente Génave (Jaén)
m10	Picual	0.31	5.70	1.69	0.12	7.21±0.5	Extra	Puente Génave (Jaén)
m11	Picual	0.20	11.60	1.82	0.12	6.75±0.5	Extra	Puente Génave (Jaén)
m12	Picual	0.22	5.50	1.72	0.12	6.67±0.5	Extra	Puente Génave (Jaén)
m13	Picual	0.18	7.50	1.78	0.13	7.36±0.5	Extra	Puente Génave (Jaén)
m14	Picual	0.28	7.20	1.64	0.12	7.05±0.5	Extra	Puente Génave (Jaén)
m15	Picual	1.28	10.46	2.15	0.19	3.3±0.5	Lampante	Lisbon

2.2. Exposure system

Figure 2 shows a schematic diagram of the automatic exposure system.

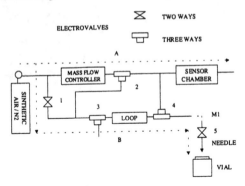

Figure 2 Automatic exposure system

A home-made exposure system, inspired in the headspace principle and adapted to the particularities of the olive oil samples, was designed and constructed. The operation of the exposure system is controlled by a measurement unit that allows the extraction of the headspace of the olive oil sample placed into the vial and drives it into the sensor chamber. The exposure system is thermo-stated to avoid condensation of the sample.

A 10 cm^3 sensor chamber was designed and built in stainless steel to avoid contamination. A volume of 1 ml of the sample is deposited in a 10 ml vial. The vial is sealed with a septum cap and it is stabilised at 40° C during 10 minutes. Up to 12 vials could be added in the exposure carousel.

The operation of the exposure system is as follows. In the first stage, the carrier gas (synthetic air 99.9990%) is divided in two pathways (A and B). An air flow of 100 ml per minute was fixed by a mass flow controller in path A, and it was passed trough the sensor chamber until the baseline is stabilised (5 minutes). This flux was the optimum for the response of the sensors. Meanwhile the needle is cleaned by the gas flowing through path B.

In the second stage, the needle is inserted (via a stepper motor) in the vial, and the vial is pressurised. The third stage starts when the vial reaches the pre-set pressure (1.5 bar). Electrovalve 3 is then switched and the pressure gradient causes the headspace of the vial to be injected into a 3 ml loop. In the final stage, electrovalves 2, 3 and 4 are switched and the airflow is passed trough the loop and conducted to the sensors chamber.

2.3. Measurement unit

The measurement unit detects the changes in the resistance of the sensors, and can also be operated to control the exposure system. It can work alone with an internal battery supply and a 2Mb SRAM non-volatile data memory, or with a computer via RS232 serial interface.

This unit is controlled by an embedded 32-bit micro-controller MC68336 working at 20 MHz and allows the control of 16 independent sensors with resistance in the range of 500Ω – 1 MΩ. Each sensor is polarized and measured independently. The polarisation circuits are described in Figure 3 and a view of the measurement unit in Figure 4.

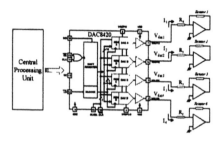 

Figure 3 Sensor polarisation circuits   **Figure 4** Measurement unit

The analog to digital conversion is made with 16 differential input 24 bit Sigma Delta A/D converter with sampling rate 10 Hz and resolution 298 nV per bit. The unit includes two outputs to interface external thermostating devices (PID) and two thermofoils for the control of the temperature in the vials and in the sensor chamber. The reading of the temperature is made with two pre-calibrated NTC thermistors with dedicated preamplifiers, with a 0.5°C accuracy and a 0.1°C resolution.

A stepper motor control interface working from 0 to 0.1 kHz regulates the insertion of the needle into the vial and the positioning of the vials. The electrovalves were controlled by open drain sink with maximum load current of 100mA and working voltage 12VDC. The system also includes an interface of a Thermo-Hygrometer via RS232C and an analog output of 0 – 5 VDC with resolution of 12 bits (1.22mV per bit) for the adjustment of the mass flow controller.

2.4. Data adquisition

Data acquisition and processing software routines were written in Labwindows CVI version 5.0 from National Instruments[TM]. The software control the measurement unit and allows representing in real time the relative change in the resistance of each sensor. The data obtained were stored in ASCII format for posterior data processing. The typical response of the sensors to an olive oil sample is shown in Figure 5.

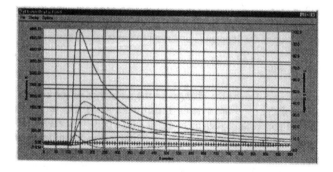

Figure 5 Response of five sensors to a sample of olive oil

Different routines were included in the software to allow data processing. In the first stage for classification of the Spanish olive oils, the height, the half-width at half-height (HWHH), the area and the derivative of each peak were calculated. Alternatively, the fast Fourier transform (FFT) formalism can be used. After several measurement of the olive oils samples a preliminary classification with this techniques was achieved.

To improve a better classification a second stage with diverse multivariate techniques was added to the software. Either the previously cited parameters of the peaks, or the first coefficients of the FFT, were used to perform a principal components analysis (PCA). Cluster analysis (CA) allowing a better separating of the olive oils in different regions. In Figure 6 a principal component analysis of five Spanish olive oils is shown.

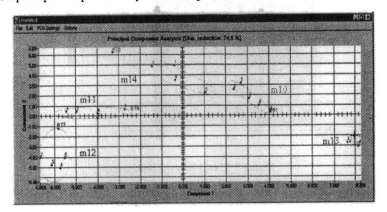

Figure 6 PCA of five Spanish olive oils

When the number of the samples was high and the clusters were overlapping, using the discriminant factor analysis (DFA), a better separation of the olive oil cluster was achieved. An example of this analysis is shown in the Figure 6.

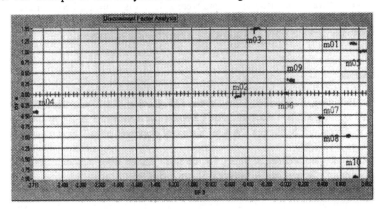

Figure 7 DFA of ten Spanish olive oil samples

The last stage of the pattern recognition software used for the classification is a neural network [6]. A multilayer perceptron network and a Fuzzy-Kohonen map were trained with the data acquired. The best performance was achieved with a 25x25 Fuzzy-Kohonen map. The neurons adapt their weights for a single olive oil, as it can be shown on the neuron map in the Figure 8. This neural network allow the classification of the Spanish olive oils under study.

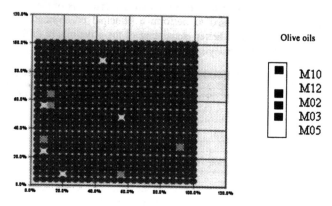

Figure 8 25xs25 Fuzzy- Kohonen map

3. Conclusions

A well organoleptically characterised olive oil samples allow the selection of a tin oxide sensors array that present the best performance in their resistance when exposed to a headspace of an aroma of olive oils. Using electronics devices for the exposure of the headspace and for the data acquisition of the response of the sensors, and advanced statistical methods via software, the array of sensors were able to discriminate between different olive oils. This artificial olfactory system built, gives rise to rapid, reversible and sensitive responses to the Spanish olive oils and allow a rigorous classification.

4. References

[1]   M. Holmberg, F. Winquist, I. Lundström, J.W. Gardner, E.L.Hines, "Identification of paper quality using a hybrid electronic nose", *Sensors and Actuators B 26-27*, pp. 246 –249, 1995

[2]   J.W. Gardner, P.N. Bartlett, "A brief history of electronic noses", *Sensors and Actuators B 18*, pp. 211. 1994

[3]   M.L.Rodríguez-Méndez, J. Souto, J.A. de Saja, "Crown ether luteium bisphthalocyanine as gas sensor", *Sensors and Actuators* B 31, pp. 1-5, 1996

[4]   M.L.Rodríguez-Méndez, J. Martinez, J. Souto, J.A. de Saja, "Lutetium Bisphthalocyanine thin films as sensors for VOCs of aroma", *Sensors and Actuators* (in press), 1998

[5]   Olías-Jiménez, F. Gutierrez, "Volatile constituents in the aroma of Virgin Olive oil IV", *Grasas y aceites* 31, pp. 391. 1980

[6]   M. Schweizer-Berberich, "Application of neural-network systems to the dynamic response or polymer based sensor arrays", *Sensors and Actuators B 26 -27*, pp. 232-236, 1995

Section 5: Applications: Food, Agricultural and Environmental
Paper presented at the Seventh International Symposium on Olfaction and Electronic Noses, July 2000

247

Application of an Electronic Nose to the Monitoring of a Bio-technological Process for Contaminated Limes Clean-Up in a Oil Rendering Plant

Corrado Di Natale, Antonella Macagnano*, Eugenio Martinelli, Christian Falconi, Emiliano Galassi, Roberto Paolesse°, and Arnaldo D'Amico

Department of Electronic Engineering, University of Rome 'Tor Vergata', via di Tor Vergata 110, 00133 Roma; Italy
* Technobiochip scarl, via della Marina, Marciana (LI); Italy
° Department of Chemical Science and Technology, University of Rome 'Tor Vergata', via della Ricerca Scientifica, 00133 Roma; Italy

Abstract. In this paper the electronic nose technology is applied, for the first time, to the monitoring of a bio-technological process used for the pollution break-down in waste limes produced in a oil rendering plant. The electronic nose here described is based on a set of metalloporphyrins coated quartz resonators. The array matched with a proper chemometrics and neural network data analysis techniques shown positive results both in qualitative terms (recognition of the different cleaning steps) and quantitative (estimation of pH, total hydrocarbons, ammonia, and phenols) of the contaminated limes.

1. Introduction.

Among the dozens of electronic noses proposed applications, the use of this technology for monitoring of industrial processes is an emerging field. In particular, the control of those industrial processes mainly concerned with dynamical evolutions that may impact with the surrounding environment are also becoming urgent.

From this point of view, oil rendering plants necessitates of a strict control both in terms of a reduction of their environmental impact than in terms of an improvement of the plant yield.

Oil refineries produces an amount of waste material in form of lime. These limes are polluted with the solid and liquid waste of the rendering processes, mainly alkanes and aromatics molecules are involved.

In the last decade, as the environmental protection urgency grew, new technologies for the cleaning of the waste have been appeared. Among them those based on bio-technologies are particularly promising. One of these methods, makes use of selected classes of micro-organisms (bacteria and fungi) able to break-down, through their metabolic action, a wide range of pollutants. In this process, known as Chemi-Osmoregulation (COR) micro-organisms are fixed on mineral carriers that can be dispersed in the polluted media.

Among the possible fixing materials the calcium-carbonate, the volcanic stones, and the alumino-potassy silicate are to be mentioned. All these substrates are porous enough to allow an optimal exposition of the micro-organisms to the polluted media.

The micro-organisms utilized for these applications usually originates from natural strains (totally saprophytic, not genetically modified).

The action of the bacteria colonies are normally monitored measuring, with a certain rate, some parameters, such as pH, Total Hydrocarbon Content (THC), ammonia, and the total phenols.

In this paper, the application of an electronic nose, based on bulk acoustic wave quartz resonators, to the monitoring of a COR process is illustrated and described.

The electronic nose, which may work on-line with the process, has shown the capability to discriminate among different steps in a COR process and to quantitative evaluate some of the chemical parameters characterizing the bacteria metabolic activity.

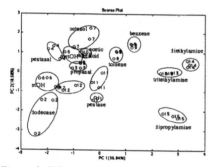 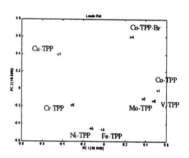

Figure 1: PCA score plot of an experiment aiming at the identification of volatile compounds

Figure 2: the loadings plot related to the scores of fig. 1. Changing the metal atom of the metalloporphyrin complex the character of the sensors changes.

2. Experimental.

Data are related to a medium size oil rendering plant in central Italy. The contaminated limes were treated with COR. Polluted limes were stocked in three different lagoons according to the development of the COR process. So that three steps in the whole process may be identified.

Data were collected over four months for a total of 86 measurements. Each measurement was performed taking a constant quantity of lime from different locations in the lagoons. Part of the lime were analyzed with conventional analytical methods in order to measure pH, THC, NH_3, and phenols. The other part of the sampled lime were analyzed by the electronic nose.

The electronic nose utilized in the experiment is one of the prototypes of the LibraNose (*University of Rome 'Tor Vergata' and Technobiochip*) [1]. The instrument is based on eight thickness shear mode quartz resonators each with a fundamental frequency of 20 MHz.

Each sensor is coated with a molecular film of pyrrolic macrocycle [2]. These molecules have been found to be particularly appealing for electronic nose applications, due to the flexibility of their synthesis process leading to molecules with broad-selectivity but oriented towards different classes of analytes [3].

An array of such molecules have been found to be able to recognize different volatile compounds. To this regards, figures 1 and 2 show the score and the loading plots respectively, of an experiment aimed at measuring different VOCs at different concentrations. In considering the different sensors contribution to the array it is worth to mention that the molecular species coating each sensor differed, one each other, for the metal atom complexed at the center of the metalloporphyrin complex.

For each electronic nose measurement, the same amount of lime was closed in sealed bottles, and held at constant temperature (30°C) for 30 minutes in order to reach an equilibrium both qualitative and quantitative) in the headspace composition.

The headspace of bottles were then fluxed into the sensor chamber. Synthetic air was used as measurement carrier, and as reference during the measurements. The difference between the steady state value of the resonance frequency of sensors in reference air under the headspace flux was considered as the sensor response feature.

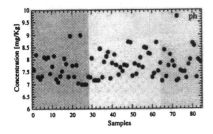

Figure 3: measured values of pH

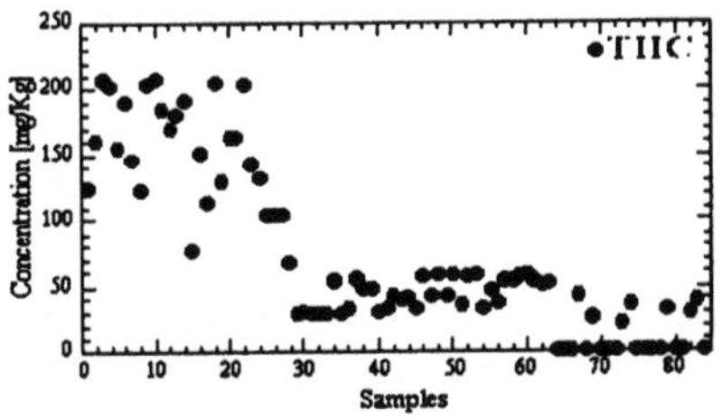

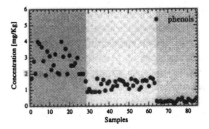

Figure 4: measured values of total hydrocarbon content

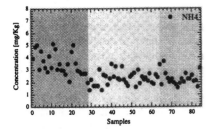

Figure 5: measured values of ammonia

Figure 6: measured values of total phenols.

3. Results and discussion.

Figure 3, 4, 5 and 6 show the values of the four chemical parameters, THC and phenols are clearly correlated with the lagoons and namely with the different steps in the cleaning-up process. Ammonia is lightly higher in the first lagoon (higher pollution), while the value of pH fluctuates without any apparent correlation with the pollution degree.

In order to emphasize the capability of the electronic nose to correctly identify among the three cleaning steps, PLS-Discriminant Analysis has been used as a data analysis method. PLS-DA allows the representation, in a suitable score plot, of the best data view in which the best class separation is achieved.

Figure 7 shows the score plot of the first two latent variables, while figure 8 shows the values of the confusion matrix in the recognition of the three classes identified in the process. As it can be argued from fig. 7, errors occurs between the first two classes , while the third class is fully identified.

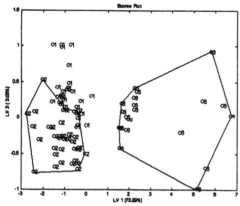

Figure 7: PLS-DA score plot of the first two latent variables, the representation is obtained looking at the best separation among the three classes.

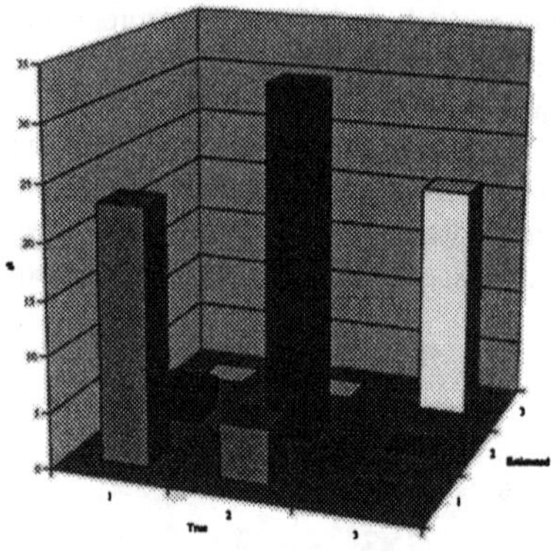

Figure 8: confusion matrix of a PLS-DA analysis of the data-set. The leave-one-out validation technique has been used.

More performing analysis has been achieved applying neural networks. As an example, a Learning Vector Quantization (LVQ) network achieved a complete identification of classes. For this scope, the data set has been divided in two sub sets, one used for training (2/3 of the total), while 1/3 has been used for validation.

An important requirement in this kind of plants is the pollution level control. For this scope the four above mentioned parameters are measured. In order to test the aptitude of the electronic nose to provide an evaluation of these parameters, a suitable neural network classifier has been utilized. For the scope, a Radial Basis Function (RBF) neural network has been applied.

Table 2 shows the average errors obtained on the validation data-set for each of the four parameters. Values below 10% have been obtained for hydrocarbon content, ammonia, and phenols. The errors related to the estimation of pH are quite large, but it has been previously shown that pH is not clearly correlated with the three cleaning steps, and definitely, with the pollution level.

Figure 9, 10, 11, and 12 show the absolute relative errors performed, for each parameter, on training and validation data. No overfitting effects are visible.

Parameter	pH	THC	NH$_3$	Phenols
Estimation error	17 %	8 %	5 %	9 %

Table 2: average errors in the estimation of the four relevant chemical parameters used to describe the pollution status of limes.

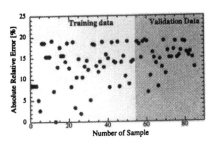

Figure 9: absolute errors for training and validation data for pH.

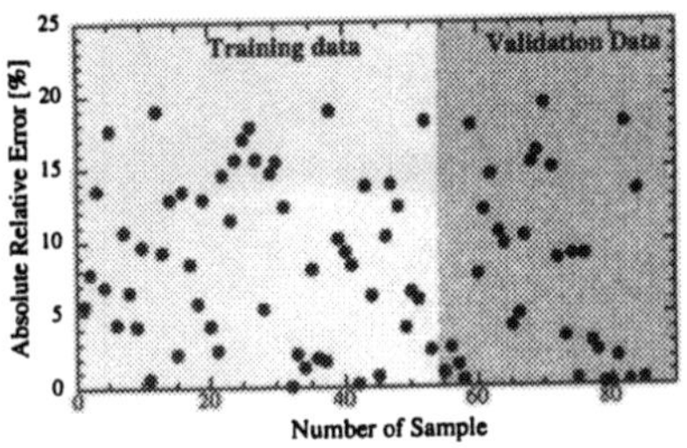

Figure 10: absolute errors for training and validation data for total hydrocarbon content.

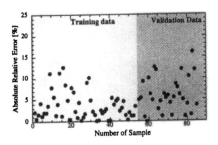

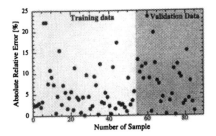

Figure 11: absolute errors for training and validation data for ammonia.

Figure12: absolute errors for training and validation data for total phenols.

4. Conclusions.

An electronic nose has been utilized to monitor a bio-technological process aimed at cleaning contaminated limes from an oil rendering plant.

Metalloporphyrins coated bulk acoustic wave sensors shown to be effective in giving good results both in qualitative terms (recognition of the different cleaning steps) and quantitative terms (evaluation of the relevant chemical parameters to describe the pollution status of the limes). Both these results have been achieved matching the electronic nose with a proper neural network based data analysis.

References.

[1]   D'Amico A., Di Natale C., Macagnano A., Davide F., Mantini A., Tarizzo E., Paolesse R., Boschi T.; Technology and tools for mimicking olfaction: status of the Rome Tor Vergata Electronic Nose. *Biosensors and Bioelectronics* **13** (1998) 711-721,

[2]   J. Brunink, C. Di Natale, F. Bungaro, F. Davide, A. D'Amico, R. Paolesse, T. Boschi, M. Faccio, G. Ferri; The application of metalloporphyrins as coating material for QMB Based chemical sensor, *Analytica Chimica Acta*, **325** (1996) 53-64

[3]   C. Di Natale, R. Paolesse, A. Macagnano, A. Mantini, C. Goletti, E. Tarizzo, A. D'Amico; Characterization and design of porphyrins-based broad selectivity chemical sensors for electronic nose applications,*Sensors and Actuators B* **52** (1998) 162-168

Section 5: Applications : Food, Agricultural and Environmental
Paper presented at the Seventh International Symposium on Olfaction and Electronic Noses, July 2000

253

Investigation of the Use of a Portable Electronic Nose Device in Truffle Industry

C N Raynaud[1], M Doumenc-Faure[1], P J Pébeyre[2] and T Talou[1]

(1) INPT- ENSCT 118 route de Narbonne 31077 Toulouse, FR
(2) PEBEYRE Ltd. 66 rue Frédéric-Suisse 46000 Cahors, FR

Abstract. The objective of this study was to investigate the ability of a portable Electronic Nose device to discriminate between truffles species which could be mixed up with Black Truffles *Tuber Melanosporum* Vitt. in Truffle marketing. Truffles clusters differentiated by Multiple Discriminant Analysis (MDA) of Electronic Nose data were related to truffles gas chromatography volatile profile. The performance of a such portable device in truffle field to locate underground truffles was simultaneously investigated. The present work showed the potential interest of using a portable Electronic Nose in Truffle industry which allowed to combine, in one single device, truffle hunting and quality control.

Introduction

Black Truffles *Tuber Melanosporum* Vitt. are underground mushrooms growing in symbiosis with especially oaks trees. This mushroom is caracterised by a very typical aroma [1]. The truffle is harvested from November until the end of March. The gathering, which requires the presence of an animal, remains a delicate and always uncertain operation. In truffle industry, the normal means of assessing truffle quality (e.g. sorted varieties) is by apparence, texture and odour assessment. Bought 'in earth' at the markets, the truffles are carefully washed and brushed before being sorted. Truffles mean *Tuber melanosporum* Vitt., this is the only variety which may be marked under the name of 'truffle' but other varieties could be confused.

The aims of this study were to investigate the ability of a portable Electronic Nose device to discriminate between truffles varieties which could be mixed up with black Truffle *Tuber melanosporum* Vitt. in truffle marketing and to evaluate the performance of a such portable device in truffle field to locate underground truffles.

Material & Methods

Three sample sets of fresh truffles (61) , *Tuber melanosporum* Vitt.(22), *Tuber brumale* Vitt.(18) and *Tuber mesentericum* Vitt. (21) were collected within the French leader company in truffle trade (Pébeyre Ltd., France) during the winter season (January-February).

Tuber brumale Vitt. (*Tb*) and *Tuber mesentericum* Vitt. (*Tmes*) varieties could be mixed up with *Tuber melanosporum* Vitt. (*Tm*). Before collection each truffle was submitted for quality evaluation by an industry trained truffle grader.

A portable Electronic Nose (PEN-2, Airsense, Germany) equipped with ten metal oxide sensors was used to measure the aroma profile of truffle samples. Before each measurement in truffle field, a pig allowed us to locate underground truffles. Results were analysed by Multiple Discriminant Analysis (MDA) and Analysis of Variance (ANOVA).

A Gas Chromatographic device DCI system (desorption-concentration-GC introduction) based on dynamic headspace concentration was used for headspace sampling before GC analysis. Truffle volatiles were analysed by GC/FID , GC/O (DN 200, Delsi Nermag Instruments, France) and GC/MS (HP 5890 serie II, Hewlett Packard, USA) [2].

Results & Discussion

MDA of Electronic Nose data illustrates scattering data points into three clusters, each containing a specific truffle variety (Figure 1). Results indicated that the ten metal oxide sensors array could detect differences between truffle variety with 84% of correct classification of samples. Canonical statistics showed that functions 1 and 2 were significant in discriminating between groups.

As the assessment occurred in industry by human nose, *Tm* and *Tb* truffles are slightly discriminated and clusters are very closed together. Variations within clusters is due to the important aroma variation between Truffles of a same variety.

ANOVA results showed that seven metal-oxide sensors were significant (p<0.05) in discriminating between truffles varieties.

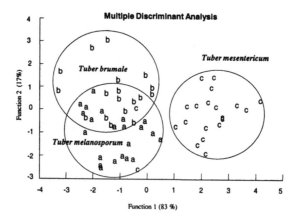

Figure 1 : Multiple Discriminant Analysis of sensor responses from different truffles varieties
a : *Tuber Melanosporum* , b : *Tuber Brumale* , c: *Tuber Mesentericum*

Table 1 and 2 show results of truffle varieties GC analysis. Methyl sulfide give an important "sulfurous" note for *Tm* and *Tb* (table 2). But alcohol chemical group (table 1) represents the most important part of the volatile fraction for *Tb* (44.3%) and *Tm* (51.1%). GC-O analysis of *Tmes* allowed to identify "musky" and "mouldy" key descriptors related to methyl anisole and ethyl anisole but ketone chemical group (25.6%) represents the most important part of the volatile fraction. The major compound identified in *Tm* was an alcohol and 2 different ketones for *Tb* and *Tmes*. These data will be useful to develop a new sensor array for truffle applications.

Table 1 : Main volatiles compound identified in fresh truffle headspace
and classified by chemical groups

Truffles Varieties	Acids	Alcohols	Aldehydes	Ketone	Sulfurous	Esters	Ethers
Tuber melanosporum Vitt.	-	51.1 (13)	18.8 (11)	15.5 (5)	12.1 (3)	0.8 (3)	1.1 (4)
Tuber brumale Vitt.	0.5 (1)	44.3 (10)	15.4 (7)	23.7 (2)	8.1 (3)	6.3 (3)	0.4 (2)
Tuber mesentericum Vitt.	1.9 (3)	18.2 (11)	19.9 (9)	25.6 (5)	7.1 (4)	-	14.7 (5)

Relative concentration % and ( ) number of molecules identified by group

Table 2 : Odorous Key Compounds and Key descriptors identified by GC-O analysis

Truffles Varieties	Odorous Key Compounds (GC-O evaluation)	Odorous Key Descriptors
Tuber melanosporum Vitt.	Methyl sulfide, 2-methylbutanal	Sulfurous, animal
Tuber bumale Vitt.	Methyl anisole, methyl sulfide	Sulfurous, musky
Tuber mesentericum Vitt.	Methyl anisole, ethyl anisole	Musky, mouldy

The overall results from experiments with the commercial portable Electronic Nose in truffle field showed that underground truffles were generally slightly detected by the PEN-2.

Figure 2 shows sensor responses at three different distances from the underground truffle. The first signal responses illustrates the sniffing of the portable EN at the same place where the truffle was located by the pig nose, the truffle was still underground. The second signal sensors response performs a measurement at the same location but when the truffle was partially discovered from earth and the third response performs a measurement when the truffle was totally discovered from earth. These results showed that underground truffles could be detected but sensors gave a very slight signal response even when the electronic nose device was very closed to the truffle.

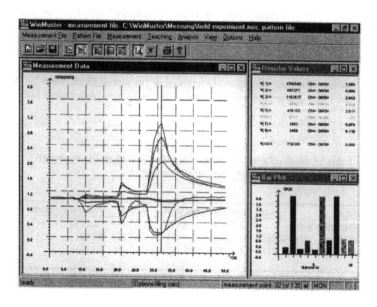

Figure 2 : Example of a measurement performed by the portable EN PEN-2 in a truffle field.

Conclusion

The present work showed the potential interest of using a portable Electronic Nose in truffle industry which could allow to combine, in one single device, truffle hunting and quality control.

Data on truffle volatile compounds is a great help to develop a specific sensor array for truffles and further developments investigating the use of a portable electronic nose as a truffle detector are actually under way both with commercial devices and lab prototype systems based on other sensor technologies in the framework of the European project EUROTRUFFE [3].

References

[1]   Talou T, Delmas M and Gaset A 1987 J. Agric. Food Chem., 35 774
[2]   Doumenc-Faure M, 2000 Doctoral Thesis, INPT (France)
[3]   Talou T, Persaud K C and Gaset A, 1992 FR Patent 2696236

Acknowledgements: This work was funded by PEBEYRE Ltd. and the European Commission under EUROTRUFFE project (FAIR 98/9556)

Section 5: Applications: Food, Agricultural and Environmental
Paper presented at the Seventh International Symposium on Olfaction and Electronic Noses, July 2000

257

Automatic Milking: an Experiment to Inspect Teats using an Electronic Nose

T.T. Mottram[1] and K.C. Persaud[2]

1 Silsoe Research Institute, Wrest Park, Silsoe, Bedford.
toby.mottram@bbsrc.ac.uk

2 Department of Instrumentation and Analytical Science, UMIST Sackville St.
Manchester.

Abstract

Methods are required for automatically detecting the hygienic condition of cows' teats before teat cups are applied by an automatic milking system. The dynamic characteristics and the odour discrimination capability of organic conducting polymers were investigated as a potential method for examining teats before milking. A controlled air flow system, an array of sensors and automated data acquisition were used to detect odours emitted by wet muck, dry muck, blood and water placed on the surface of an artificial teat. Canonical variate analysis of the output of 20 sensor elements subjected to a total of 28 random replications of the experimental procedure was able to discriminate between four treatments and the control ($P = 0.1$). Further work is required to adapt this laboratory system to detect teat contamination in field conditions.

1. Introduction

Systems have been developed to automate many of the functions of the herdsperson in the milking parlour (Ipema et al., 1992). However, before automatic milking can achieve the same level of quality control as the human operator, methods of teat inspection and cleaning will have to be developed. Mottram (1992) proposed that an electronic odour detection system could be used to inspect the teat for signs of dirt or injury.

The main contaminant on cows' teats is fresh bovine faecal material (muck) splashing from floors or picked up from beds. This can be either wet, or dry if it has adhered for a few minutes. Some teats suffer from wounds and lesions due to trampling between milkings. Milk hygiene regulations insist that teats with wounds must not be milked because to touch these may be painful to the cow and may lead to contamination of the milk. Wounds may be detectable by the odour of blood on the surface of the teat. The milking system provides a short interval after the cow has entered the stall in which to inspect and clean the teats before milking. This time interval is limited by the neuro-hormonal control of the cow's milk ejection reflex. Failure to extract milk within the reflex period leads to reduced milk yields (Hamann and Dodd, 1992). It is essential that an inspection system should make a decision about the need to clean or milk the teats within a few seconds of the cow entering the milking system.

Prototype sensing systems based on image processing (Bull et al., 1996) and optical analysis (Bull et al., 1995) have been tested with limited success. An alternative approach would be to use an array of sensors to detect odours evaporating from contaminants on the surface. Conducting polymer sensors have been developed at UMIST (Persaud et al., 1991,1992) and at other centres (Bartlett and Gardner, 1992). Films of conducting polymers consist of polycationic chains of aromatic or heteroaromatic molecules, with the positive charges counterbalanced by anions. The conductivity of these materials may be modulated reversibly when various chemicals are adsorbed, the sensitivity being generally greater for polar molecules. Specificities of individual sensors to groups of chemicals are relatively broad, but the selectivity of individual sensor elements can be tailored by adapting the chemical structure of the polymers to the type of chemical groups to be detected. Hence an array of sensors, when exposed simultaneously to the identical concentration of a volatile chemical produces a pattern of responses that may be used as a descriptor of the chemical species adsorbed.

The object of this work was to determine the feasibility of using conducting polymers as sensors of teat cleanliness in a model milking system. The features examined were time of response to a change in odour and discrimination between different contaminants.

2. Materials and methods

The apparatus is illustrated in Figure 1. Ambient air was drawn either through a bypass pipe or through a rigid teat cup containing an artificial teat of polytetrafluoroethylene (ptfe) of diameter 20 mm. The air was then drawn by vacuum over an array of 20 conducting polymers at 320 ml/min. The array consisted of 20 sensor elements, based on substituted polypyrroles[9], each being 250 x 250 μm, arranged on the perimeter of a square ceramic chip carrier of width 10 mm. The relative humidity of the input air was adjusted to approximately 60 % by mixing air from two sources, one ambient and one saturated. Input air flow and relative humidity were controlled manually.

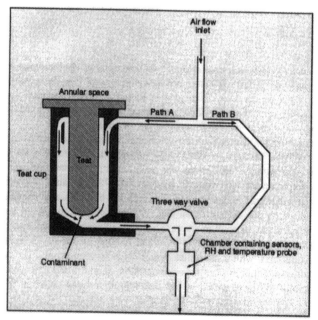

Figure 1. Schematic diagram of apparatus used to test the potential for detecting contaminants on an artificial teat with an electronic nose

Contaminants on the surface of the teat were simulated by soaking a 100 mm^2 segment of paper tissue with contaminant and attaching it by surface tension to the teat end. The contaminant treatments were wet muck, dry muck, blood, distilled water and a clean teat as a control. Muck was gathered from a fresh bovine faecal deposit and stored at 4^{0}C overnight. Bovine blood stored with sodium citrate as an anti-coagulant was used to simulate a wound. Since additives to blood are heavily buffered, the sodium citrate had no odour effect. The dry muck treatment was achieved by attaching a segment of paper tissue with wet muck and drying it in situ. All surfaces in contact with the sample air were made from ptfe. The teat and apparatus was washed with distilled water between each treatment and allowed to dry. The order of treatments was randomised to minimise the effect of any residual odour (either adsorbed onto the sensor or apparatus) on the measurements.

The circuit for each sensor (Persaud and Travers, 1991) was such that the sensor formed the feedback loop of a current to voltage operational amplifier. The output of this amplifier was directly proportional to the base resistance of the sensor. The base resistance value was output from the digital to analogue converter (DAC) to a differential amplifier, the output of which was connected to a programmable gain amplifier. For each sensor, a base resistance and gain setting were stored in memory during a calibration phase and used each time the sensor was accessed via the multiplexor.

Each sampling run started with humidified ambient air being passed by path B (Figure 1) through the sensing chamber for 120 s to calibrate the sensor by establishing the base resistance. The sensing run then began. The three-way valve was switched 10 s after the calibration phase so that air passed over the teat with its attached contaminant (path A, Figure 1) and then through the sensing chamber. The valve was switched back 30 s later to path B. The sensor values were recorded for a further 80 s.

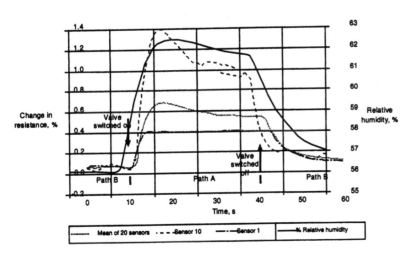

Figure 2. Typical sensor responses during the cycle of exposure to ambient, contaminated and ambient air passing through the sensor chamber.

The change in resistance of each sensor was recorded at 1 s intervals. The base resistance (R) for each sensor in the array was taken as the mean resistance during the calibration period. The change of resistance, dR, was divided by R for each sensor.

The typical response of the sensor system during the sensing phase is shown in Figure 2. After the valve was switched, a rapid rise in dR/R to full scale output occurred. The responses of sensors varied, sensor 10 in this example exhibited the greatest response whilst sensor 1 exhibited the least. The steady state output was calculated as the mean value for the period from 15-30 s after the valve was switched to path A. The valve was switched to path B so that clean air passed over the sensors, causing resistance values to drop towards the original base value. The profile of the mean dR/R values was similar to that of the relative humidity which tended to rise within a range from 55% - 70% when air was passed over the wet samples.

The response was measured as the fractional change of resistance occurring when an odour was presented to the sensor. This was given by the dR/R where R was the base resistance of the sensor in ambient air at the start of each sample period, and dR the change in resistance. To compress the amount of data collected during the sensing run (120 s duration) to a single point, dR was defined as the maximum change in resistance (In the example shown in Figure 3, t=17 s). The order of presentation of samples was randomised and clean air blown through the apparatus for a minimum of 5 minutes to remove lingering odours. The data set matrix comprised 28 columns of the treatments by 20 rows of the maximum sensor values recorded during the sensor run on the relevant treatment. The data set matrix was then analysed using canonical variate analysis (Manly, 1994). This process calculated sums of squares and the variances between the groups of samples (5 clean, 6 wet muck, 7 dry muck, 6 water, 4 blood) and the sums of squares and the variances within groups.

Visual inspection of multivariate data is inherently difficult. This method of analysis allowed reduction of multidimensional data to two dimensions, clarifying visual inspection of the separation between groups. The axes represent the rotated eigenvectors in multidimensional space and are without units. The variation over all the groups was compared with the within-group variation. The comparative variation of the distances can be represented on a two dimensional plot.

3. Results

Table 1 records the times between switching the valve to path A and the sum of the resistances reaching a maximum; together with the time between switching the valve, back to path B and the output falling to below 20% and 10 % of the steady state output. As expected, the output from the control treatment was small and hence the same analysis could not be conducted. In the canonical variate analysis the clean teat values were taken as the values for the sensors (dR/R) at 12 s (which was the mean of all recorded groups from Table 1).

Table 1. Response times (s) for $\sum dR/R$

	Time (s) from zero to maximum response		Time from valve off to 20% of steady state ouput		Time from valve off to 10% of steady state output	
	Mean	SD	Mean	SD	Mean	SD
Blood	12.2	2.4	8.6	2.1	14.2	3.5
Dry muck	15.3	4.3	9.0	1.2	18.6	3.5
Moisture	10.8	3.6	7.3	0.7	13.2	2.9
Wet muck	7.8	1.9	8.8	1.9	17.8	6.6
All	12.0	4.3	8.4	1.6	16.0	4.7

The canonical variation of the dR/R values is shown in Figure 3. The grouping of the points within the circles described by a radius of $\sqrt{\chi^2}$ with two degrees of freedom (the Mahanalobis distances) from the centroid of area of the class indicate that these results are not random $P=0.1$. These data indicate that the sensor array could discriminate between the different contaminants of teats.

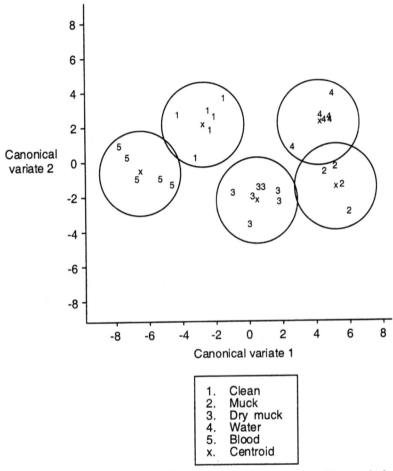

Figure 3. Discrimination between different samples of contaminant with canonical variate analysis of the peak resistances of the sensors when exposed.

4. Discussion

The sensor system could discriminate between the contaminants and within a maximum of 22 s from the change from clean air to contaminated achieved 100 % full scale output (illustrated in Table 1 and Figure 2). Control treatments (clean teats) gave much lower sensor response amplitudes than the other treatments, and it was not possible to define a maximum response for use in Table 1, although a representative value was used   for   the   canonical variate analysis. Control treatments should not have differed from the calibration phase of the experimental run. However, the maximum summed absolute response for the control was Σ dR| /R=0.3 % compared with 0.5-1.1 % for treatments. It is possible that some residual odour in the apparatus was responsible for the change in signal.

The rates of adsorption (column 1, Table 1) and desorption (columns 2 and 3) of odour compounds by the sensors differed (Table 1). The adsorption phase overshot the steady state value in all cases whereas the desorption phase rarely did so. This may have been owing to the development of a concentration of odour in the teat cup before the valve was switched to pass air over the teat. on switching, this would have caused a higher concentration of odour to arrive at the sensor to be replaced after a few seconds by a steady emission of odour from the sample.

The threshold of 10% of the steady state output was arbitrarily selected as being a level below which all sensors returned, but sometimes the sensor array took several minutes to return to its base value. Irregularities in the rates of adsorption and desorption need further investigation. Techniques for calibrating sensor systems to low odour levels need some attention. Methods have been proposed to develop analogue odours which can be cross related to measurement by precise techniques such as gas-chromatograph mass spectrometry (Elliott-Martin et al., 1995).

The change in conductance observed when molecules are adsorbed on the surface of the materials is dependent on the affinity of binding as well as the vapour concentration. Hence other molecules such as water may compete for binding or displace adsorbed molecules from the conducting polymers. This generally results in a loss of sensitivity of the sensor at high relative humidities. In these experiments the humidity of the ambient input air was held approximately constant (at the level that would be expected near a cows' teat). Although the intensity of the signal recorded would have changed if sample humidity varied, this did not affect odour discrimination as shown in Figure 3.

While the sensors have been shown to work in a laboratory situation, in practice the variability of ambient humidity the difficulties of controlling incoming airflow requires a re-engineered sampling system. The sensors could also be deployed so as to draw a sample from air above the milk as it flowed through the milking system. This might allow other parameters such as cow health to be monitored.

5. Conclusions

Conducting polymer odour sensors can discriminate between different contaminants on a teat within a teat cup. To develop a practical device, methods of controlling the effect of ambient humidity and incoming airflow on the sensors will be necessary.

Acknowledgements
This work was partly funded by the Ministry of Agriculture, Fisheries and Food, Great Westminster House, London.

References
Ipema A H   Lippus A C   Metz J H M   and Rossing W   (Editors) 1992 Proceedings of the International Symposium Prospects for Automatic Milking   Wageningen   Netherlands   23-25 November 1992 (EAAP Publication No.65 1992) Pub. Pudoc Wageningen Netherlands

Mottram T T   1992 Teat Inspection Device (Olfactory) GB Patent Application 9224404.5

Hamann J and Dodd F H   1992 Milking Routines In: Machine Milking and Lactation Ed: Bramley A J Dodd F H Mein G A and Bramley J A   Pub: Insight Books Newbury Berks UK pp 69-96

Bull C R   McFarlane N J B Zwiggelaar R Allen C J and Mottram T T 1996 Inspection of cows teats by colour image analysis for automatic milking Computers and Electronics in Agriculture 15 15-26

Bull C R   Mottram T T and Wheeler H   1995 Optical Inspection of teat cleanliness for automatic milking Computers and Electronics in Agriculture 12: 121-130

Persaud K C and Pelosi P 1992 Sensor arrays using conducting polymers in: Sensors and Sensory Systems for an Electronic Nose Gardner J W and Bartlett P N (Editors) Kluwer Academic Pub pp 237-256

Persaud K C and Travers P (1991) Multielement arrays for sensing volatile chemicals Intelligent Instruments and Computers 147: 147-54

Bartlett P N and Gardner J W   1992 Sensors and Sensory Systems for an Electronic Nose (Editors) Kluwer Academic Pub.

Persaud K C and Travers P J 1997 Arrays of Broad Specificity Films for Sensing Volatile Chemicals In: Handbook of Biosensors and Electronic Noses: Medicine Food and the Environment Ed E Kress-Rogers CRC Press New York pp 563592

Manly B F J 1994 Multivariate Statistical Methods Pub: Chapman and Hall London

Elliott-Martin R J   Bartlett P N   Gardner J W and Mottram T T 1995 An overview of electronic noses and their applications In: Sensors and their applications VII Ed. A T Augousti pub. Institute of Physics Bristol UK

Analysis of Off-Flavours in Raw Cow's Milk with a Commercial Gas-Sensor System

J.E. Haugen, O. Tomic, F. Lundby, Knut Kvaal[#], E. Strand*, L. Svela*, and K. Jørgensen*

MATFORSK, Norwegian Institute for Food Research, Osloveien 1, 1430 Ås,[#] Agricultural University of Norway, Dept. of Mathematical Sciences, Post box 5035 N-1430 Ås, *TINE, Norwegian Dairies, Centre for R&D, Post Box 50, 4358 Kleppe

Abstract. A commercial chemical sensor array system has been applied to analyse fresh milk from dairy cows with different off-flavours in parallel with sensory assessment. It was demonstrated that the standardised method could be used for the classification of milk with and without feed-off flavours. Using milk reference samples for calibration in combination with drift compensation algorithms showed that it is possible to maintain a satisfactory long-term reproducibility of the sensor array system.

1. Introduction

There is a need in the dairy industry for rapid techniques for quality control of raw material, processing and final product. Of crucial importance is the quality of the raw milk used for production. The chemical and microbiological control of milk in Norway today is to a large extent based on automated analysis systems, but the sensory control is based on sensory assessment by a trained panel of unpasteurised farm milk. This quality control scheme is used for excluding milk that does not meet the standards from the market, and as a basis for the payment and production control for the farmers. This sensory assessment is a time consuming method, and a method that is difficult to standardise. Also, there is a risk for exposing the participants of the taste panels for infection from pathogenic bacteria in the milk.

Today there is no existing method that fully can replace the sensory assessment in quality control of raw milk. Consequently, there is a need for development of techniques that can identify milk with off-flavour in a rapid, cost efficient and reproducible way.

The dominant classes of compounds related to milk flavour and off-flavour make up esters, sulphur compounds and aldehydes [1-2]. Since commercial chemical sensors are sensitive to these compounds chemical sensor technology should in principle also be applicable to determination of quality of unpasteurised milk. Few chemical sensor-array studies on milk have been reported in the literature, and this has been mainly on milk quality related to pasteurised milk [3-5].

The objective of this project has been twofold: To obtain a standardised reproducible gas sensor analysis method for raw milk. And to evaluate the feasibility of a commercial gas-sensor array to discriminate raw cow's milk with different off-flavours from normal (good) milk samples according to a routine sensory assessment procedure.

2. Experimental

2.1. *Sample material*

250 samples of different flavour quality that had been assessed by the dairy sensory panel were analysed. The off-flavour sample categories measured were respectively lipolytic rancid (38 samples) feed (92 samples), malt (8 samples), oxidised (20 samples), salt (10 samples) and sour (2 samples). In addition, 80 samples were analysed that were assessed as good without off-flavour by the sensory panel. Fat content of the measured samples was in the range 3.5 –4.5 %.

The measurements were performed in parallel with sensory assessment. On-site measurements were carried out over a 3 months period at two different dairy laboratories with two different trained sensory panels using a standardised procedure for tasting. A commercial gas-sensor system (NST 3320) consisting of 10 MOSFET and 12 MOS sensors was used to analyse the milk samples. 5 ml milk samples were placed in 30ml clear glass bottles and sealed with pre-heated silicon septa's and plastic screw caps. The samples were incubated for 20 min. at 60 °C. Headspace gas was purged into the sensor system with a flow of 30 ml/min for 50 sec., i.e.25 ml headspace pr. sample. About 40 samples were analysed daily together with calibration samples.

2.2. *Calibration samples and data transfer*

Artificially oxidised milk samples were used as calibration samples. Aliquots of 100 ml samples were kept frozen at -80°C and thawed prior to analysis for every measurement experiment. Two calibration samples were used for each measurement sequences, i.e. total 60 calibration measurements over 30 sequences during the three months measurement period. In cases where shifts in the response of the calibration samples were observed between measurement sequences or within a sequence a drift compensation and shift algorithm was applied to compensate for these changes [6].

2.3 *Data analysis*

The signal height of the transient relative to the baseline signal was used as raw data for the data processing. The NST system software Senstool and The Unscrambler (CAMO version 7.5, Trondheim, Norway) were used for the principal component analysis and statistical evaluation. Three different classification methods were used for classification and prediction of off-flavours; discriminant partial least squares regression (DPLSR), SIMCA classification (The Unscrambler 7.5.) and an artificial neural network (ANN) classification routine constructed as outlined in [7]. For a more extensive explanation of the Backpropagation network we refer to [8].

3. Results and discussion

3.1. *Discrimination of off-flavours*

Due to drift in the sensor signal over the three months of measurement, the drift compensation algorithm [6] was applied. The major drift occurred between the different days of measurement but in some cases a significant drift was also observed within single sequences and it was required to apply the drift compensation method within single measurement sequences. After applying the drift compensation algorithm based on the milk reference samples, the day to day signal shifts in sensor responses were eliminated and real features in the measurement data was preserved. This allowed the transferring of all the measurement data obtained over three months into one data file for the proceeding data analysis.

The sensor responses of the MOS sensors in general showed higher responses for nearly all feed off-flavoured samples compared to the other samples including also the good accepted samples (Fig 1). No systematic distribution in sensor response for different off-flavours could be demonstrated for the MOSFET sensors.

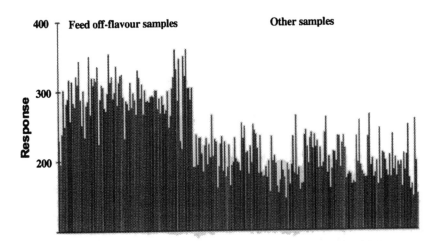

Figure 1. Typical sensor response (mV) distribution for one of the MOS sensors for all the samples measured.

A PCA plot based on the responses from 8 drift corrected MOS sensors are displayed in (Fig 2). A high explained variance (99%) is described by the two first components indicating that there is not much more systematic variation left in the higher components. The sensor responses used in this data analysis were 8 selected MOS sensors that showed the greatest discrimination ability. It is seen that most of the feed off-flavoured samples are grouped to the right along the first axis and that there is a slight overlap in the centre with the other samples. This distribution was also confirmed by inspecting the single sensor responses (Fig.1).

All the remaining samples including the good samples are randomly grouped to the left with no clear discrimination of either of the flavour categories, except for the two sour samples far below the main clustering. Accordingly, no significant difference in vapour composition was detected by the sensors for these off-flavours (rancid, oxidised, malt, salt) indicating that they have a similar headspace composition and also that they by the applied measurement conditions seem to be similar to good samples. A significant variation and overlap of the lipolytic rancid, oxidised, salt and malt flavours could possibly be ascribed to some uncertainty in the sensory assessment over the measurement period.

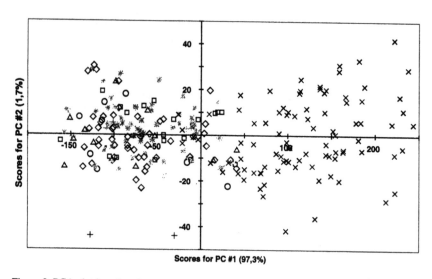

Figure 2. PCA plot based on 8 sensor responses. Symbols: sour (+), feed (x), malt (o), rancid (◊), oxidised (□), salt (Δ), good (∗).

Feed flavour in cow's milk is a flavour defect that is related to the feeding of cows. Dimethyl sulphide and volatile aldehydes, ketones, alcohols and esters are typical compounds that are released from dairy feed like silage and green forages [9]. The few feed related samples that grouped to the left of the second principal axis where samples that also had received a remark by the sensory assessment on lipolytic rancid or oxidised, which was in accordance with the grouping of the lipolytic rancid and oxidised flavoured samples.

The two sour samples are clearly separated from the other samples at the lower left in the PCA plot (Fig. 2) suggesting that the electronic nose should have a potential to point out this flavour defect in cow's milk. The flavour defect is caused by the growth of lactic acid producing bacteria when samples have been exposed to elevated temperatures [2]. This flavour is very easily perceived by sensory assessment since it represents a very characteristic sour odour by the volatile acids. However, the limited number of samples makes it difficult to make any conclusive statement to whether the sensor array system discriminates this particular flavour defect. More measurements of samples with this flavour defect need therefore to be analysed. The malt off-flavour is related to bacterial growth of *Streptococcus lactis* var. *maltigenes* when the milk has been insufficiently cooled and results in the formation of volatile branched-chain aldehydes [11].

The lipolytic rancid off-flavour is related to increased level of free fatty acids. It has previously been suggested that the volatile free fatty acids are not volatile enough and partly their polarity prevents them from being drawn into the headspace vapour of milk to obtain significant concentration levels in the vapour phase [10]. However, this will be strongly dependent on sampling conditions. In the reported case the sampled headspace volumes after preheating of samples was around 200 ml. In our case an only 25 ml headspace volume was analysed. This may explain why no discrimination could be obtained for these samples.

The oxidised flavour is related to the presence of volatile aldehydes and ketones that are derived from oxidative breakdown of unsaturated fatty acids [12]. These volatiles occur at sub ppm to ppm levels in the vapour phase and should in principle be within the sensitivity range of the sensor system.

The poor discrimination of the off-flavours except from the feed and sour flavour samples and the fact that they could not be discriminated from good samples suggest that the volatiles related to these kind of off-flavours were not significantly present as major components in the vapour phase. It could as well indicate that the sensors used do not have the selectivity and sensitivity that would be required to point out these flavours. Altering the sampling conditions by enhancing sampling volume could possibly reveal whether there is more to gain in discrimination of these of samples. The finding that the salt flavour could not be discriminated was not surprising, since this off-flavour will basically be related to components entirely present in the liquid phase.

3.2. Classification

According to the outcome of the PCA plot (Fig. 1) using responses from 8 of the sensors, the data set was divided into two major classes of samples: One class containing all the samples with feed off-flavour and a second class with all the remaining samples. The two sour samples were excluded from the data due to the poor modelling ability of the limited number of samples to build a reliable PCA class model. The whole data set (250 measurements) was split into a training set and a validation set with 50 % of the samples randomised in each data set.

For the SIMCA classification separate PCA models were calculated for the feed-samples and the non-feed samples respectively. A 3 principal component model (99 % explained variance) was used for both models. The model to model distance between the two classes was 12, showing that the two classes were significantly different. However, the classification results showed that there was a significant overlap of samples (Table 1).

For the DPLSR classification the numeric value 1 was used as a classifier variable for the feed class and 0 for the remaining non-feed samples. The treshold criterium for belonging to a class was set to 0.5 i.e., samples with prediction values below 0.5 were classified as non-feed samples and those exceeding 0.5 were classified as feed samples. The two first principal components of the PLSR model were significant and were used for the predictions.

Method	Classification rate feed off-flavour	Classification rate other samples	Total classification rate
SIMCA	80.4	55.7	64.8
DPLSR	91.3	98.7	96.0
ANN	90.3	100.0	96.7

Table 1. Results from classification modelling of samples with feed off-flavour and samples without feed flavour. The classification rate is given in % of the number of samples that classified correctly to its respective class. Total classification rate refers to the total number of samples classified correctly to its respective classes.

For the ANN classification a non-linear feed forward network with the backpropagation learning rule was constructed where the 8 sensors were used as inputs and 2 hidden neurones were found to be optimal. A cross-validation optimisation was used to estimate these parameters. The two classes were represented by two binary variables representing the respective classes with and without feed off-flavour. A sigmoid transfer function was used as the non-linear transfer function. Predictions of the classes were represented by the actual values rounded to 1.

DPLSR and ANN performed equally well and indicate a linear relationship in the data (Table 1). The poor outcome of the SIMCA classification has to do with the fact that the major variance in the whole data set is along the first principal component (Fig 2.). Accordingly, the object class distance along the second dimension would be similar for the two PCA class models and suggests that SIMCA in this case will not be a proper method for classification.

The DPLSR aims at separating the classes by specificly finding the variance in X that discriminates the classes. SIMCA models the class variations, which may express other types of systematic variation in the X-data. When significant variables from a DPLSR is applied in SIMCA models, this should yield similar model performance as for the DPLSR. On the other hand, the critical limit for DPLSR is rather ad hoc in contrast to the more formal statistical based critical limits in SIMCA.

The simple neural network approach used may easily be implemented as a classifier in on-line applications. The prediction calculation time is short. When the responses are used as input to the neural net, it is verified that a simple network is sufficient to achieve a good performance.

4. Conclusion

A standardised gas-sensor array method for the classification of farm milk in order to compare results and improve the long-term reproducibility of the method has been developed. This has been achieved by using a standardised reference milk sample for calibration in combination with drift compensation algorithms. The results show that a commercial gas-sensor array is able to discriminate between raw farm milk with- or without feed off-flavour.

The sour off-flavour could also be discriminated, but the limited number of sour samples was not sufficient to justify this conclusion. The finding that the off-flavours oxidised and malt could not be discriminated could possibly be ascribed to the chosen sampling conditions. More work is needed on these off-flavours to see whether they could be pointed out with a chemical sensor array system.

References

[1]   Moio L, Dekimpe J, Etievant P and Addea F 1993 J. Dairy Res. 60 199-213
[2]   Badings H T 1991 In H. Maarse (Ed.) Volatile compounds in Foods and Beverages 91-106
[3]   Korer F, Diego A, Balaban L and Balaban L 1999 In: *Electronic nose & sensor array based systems*, Technomic Publ. 154-162.
[4]   de la Pinnsonnais M and Innings F 1998 Electronic Noses in the Food Industry, 16-17. Nov. Stockholm 42-48
[5]   Haugen J E, Kvaal K and Rødbotten M 1997: Sem. Food Anal. 2 207-214.
[6]   Haugen J E, Tomic O and Kvaal. K 2000 Anal. Chim. Acta 407 23-39
[7]   Haugen J E and Kvaal K 1998 J. Meat Sci. 49 273-286

[8]   Fauset L 1994. Fundamentals of Neural Network. Architectures, algorithms and applications. Prentice Hall 289-333

[9]   Gordon D T and Morgan M E 1972 J. Dair. Sci. 55(7) 905-912

[10]   Vallejo-Cordoba B and Nakai S 1993 J. Agric. Food Chem. 41 2378-2384

[11]   Morgan M E 1976 Biotechn. Bioeng. 18 953-965

[12]   Forss D A, Pont E G and Stark W 1955 J. Dair.Res. 22 91-102

Section 5: Applications: Food, Agricultural and Environmental
Paper presented at the Seventh International Symposium on Olfaction and Electronic Noses, July 2000

273

Monitoring of Odour Emissions from Agriculture with Chemical Sensor Arrays

Barbara Maier, Gisbert Riess

Bayerische Landesanstalt für Landtechnik, TU München, 85350 Freising, Germany
Phone: ++49 8161 715341, Fax: ++49 8161 714363 email: maierb@tec.agrar.tu-muenchen.de

Abstract: Odour emissions of agricultural housing systems were rarely recorded, because there is no continuous measurement method available. A new method based on a multisensor array was introduced, which can record the odour emissions continuously. A multisensor array with semiconductor sensors was used, which was calibrated with the olfactometry. The emissions of different housing systems for fattening pigs were compared by this system. Due to the continuous measurement it is possible to evaluate the housing systems even in night time and on weekends. This always is a problem when using an organoleptic panel. Concerning the odour emissions the conventional heat insulated system was compared to new up coming outdoor climate stable systems.

1. Introduction

So far, olfactometry has been the only available technique for odour measurements. In this case, the human nose serves as a sensor, i.e. test persons are employed as sniffers. Due to its dependence on test persons, and working hours this method has disadvantages. It is not possible to measure the emissions continuously, which is necessary to assess different agriculture systems. The reason is that odour nuisance often is a problem specially at night times or on weekends. Therefore it is necessary to develop a technical measuring device for continuous odour monitoring that is independent from the human nose. A new system based on a multisensor array was developed for continuous measurements.

2. Experimental

Our experiments were done on a farm where our institute had a long term experiment in gas monitoring run [1]. For odour monitoring the multisensor array FOX 4000 from Alphamos was used. This system has 18 semiconductor sensors. For use outside the laboratory, the whole sensor system including the humidification device was put in a heat insulated box. This outdoor system was located on a central point of the farm.

Three measuring points inside tree different housing systems where installed. From these points the stable air was pumped through heated gas tubes to the sensor array. On line measurements were made each 20 minutes. There was round about an hour between measurements from the same measuring point. Meanwhile the sensor monitoring on the farm, samples of stable air were taken twice a day into Tedlar bags. These samples were used for olfactometric concentration measurements [2]. A trained organoleptic panel of at least 4 persons detected the odour concentration.

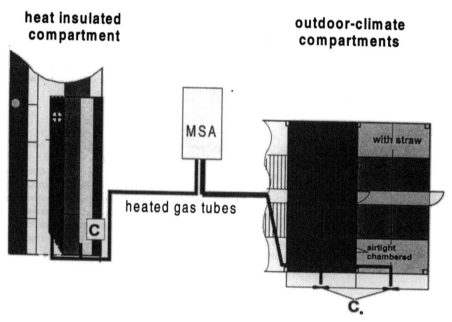

Figure 1: Schematic overview of the farm

The experiment took place in a new naturally ventilated building with two different compartments. The one experimental compartment contained kennels and partly slatted floor, the other kennels and a bedded area with 100 to 200 g straw per pig per day. In each compartment 64 animals were hold. The lying-area for the pigs was in the heat insulated kennel. Feeding in all stables was realised by mash feeders. In the experimental compartment with partly slatted floor the slurry was stored under the slats for about 6 to 8 weeks, in the experimental compartment with straw, the liquid manure was also stored under the bedded area, the dung was removed out of the pens every two to three days.

The third experimental compartment was also located on the same farm in a heat insulated stable with partly slatted floor and forced ventilation. Incoming air was led through a channel above the slatted area, outgoing air left the compartments by a ventilator.

The measuring of the air flow through the naturally ventilated experimental compartments is done by a further developed method of the "large dynamic chamber". The two experimental compartments were separated air tightly from each other and from the remaining stable. A temperature control is installed in the remaining part of the stable, detecting the temperature continuously and comparing it with the temperature in the experimental compartments. The measurement of the air flow out of the heat insulated system is also done. So the air volume through the experimental compartments can be measured exactly.

3. Results

A relation between the odour concentrations measured by the organoleptic panel and the MOX sensors was found as follows.

$$y = 24{,}233x - 41{,}97$$
y: odour concentration
x: sum of the sensor signals

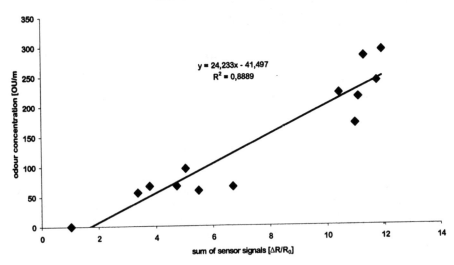

Figure 2: Calibration function for the sensor signals

According to this equation the MOX sensor array was calibrated by the olfactometry [3-6]. This can be interpreted as correlation between the substances that can be measured by the non specific sensors and the selective receptors in the human nose [4]. Such relation must be done for each specific application again.

The odour concentrations were continuously monitored by the multisensor array for 10 days based on this calibration. The following figure shows the odour concentrations in this period:

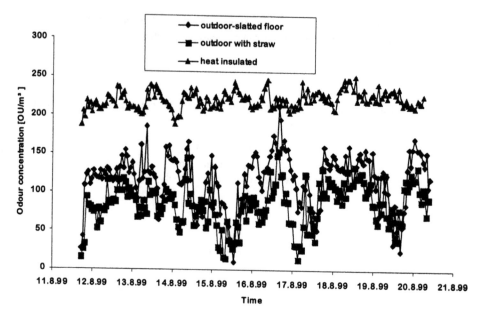

Figure 3: Odour concentrations in three different housing systems for fattening pigs

As the figure 3 shows there are lower concentrations in the outdoor climate stable compartments. The compartment with straw is a little lower than the compartment without straw. There is a big fluctuation in the concentration which is caused by fast changing wind speed and wind directions inside the outdoor stables. The heat insulated stable has the highest concentrations during the whole measuring period. Inside this stable there is little variation in odour concentrations. These stable conditions are mainly caused by the isolation and the forced ventilation.

The problem of odour nuisance from agriculture is not solved when monitoring the odour concentrations. The neighbourhood is less interested in concentrations inside the stables than in the odour emission rate. The ventilation rate was documented continuously as described in the method, so that the emission rates could be calculated. In order to make the comparison of different housing systems easier the emission rate is standardised to 500kg Life weight of the animals. The emission rate is the result of odour concentration multiplied with the ventilation rate divided through the standardised life weight.

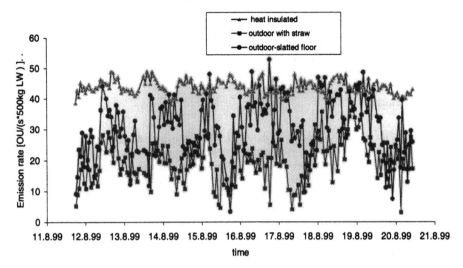

Figure 4: Odour emission rates of three different housing systems for fattening pigs

As already documented in the concentrations also the emission rates printed in figure 4 show similar results. The outdoor climate stable compartments have lower emission rates than the heat insulated stable. The fluctuation in emission rates at the outdoor climate stable compartments is higher. According to the different flow conditions of the wind the emission rates differ strongly. The mean of the emission rates that come from the outdoor climate stables is lower. The heat insulated stable has the highest emission rates during the whole measuring period.

For neighbourhood this means that with the new housing systems with outdoor climate less odour nuisance could be expected.

4. Discussion

The figure shows that the odour concentration fluctuates strongly. The housing system outdoor climate with straw has the lowest odour concentration followed by the other outdoor housing system and the heat insulated housing system. This demonstrated that the housing system outdoor climate with straw leads to low odour concentrations compared to the other housings systems.

This example demonstrates that the multisensor array in combination with the olfactometry is able to record odour emissions continuously and improves the existing techniques.

This is a practicable procedure for odour monitoring of complex gas mixtures. When continuous calibration of the sensor array is done using the organoleptic panel there is no problem with regard to sensor drift expected. Even if sensors are replaced this has little effect on the measurements.

The use of multisensor array techniques outside the laboratory was managed successful. A new approach was made to environmental monitoring with regards to volatile components or even odour monitoring.

5. Acknowledgement

This work has been supported by the Bayerischen Staatsministerium für Landesentwicklung und Umweltfragen (BayStMLUF).

6. References

1. Stegbauer, B.; Neser, S.; Gronauer, A. ;Schön, H. (1999): Vergleich der Emissionen klima- und umweltrelevanter Gase aus verschiedenen Mastschweinehaltungssystemen - Konventioneller Vollspaltenstall und zwei Außenklimastallvarianten. In: Bau, Technik und Umwelt in der landwirtschaftlichen Nutztierhaltung. Tagungsband zur 4. Internationalen Tagung in Freising, 9.-10.3.1999. Hrsg.: Institut für Landtechnik der TU München-Weihenstephan. Münster: LV Druck im Landwirtschaftsverlag GmbH, S. 87-92 (1999).
2. VDI 3881 Bl. 1-4: Olfaktometrie, Geruchsschwellenbestimmung-Grundlagen. Beuth-Verlag, Berlin (1986)
3. Persaud, K.C.; Khaffat, S. M.; Hobbs, P.J.; Sneath, R. W. Chemical Sense 21,495-505 (1996)
4. Maier B., Riess G., Gronauer A.; Landtechnik 55-1, S. 44-45 (2000)
5. Maier B., Riess G., Gronauer A.; Agrartechnische Forschung 6-1 (2000), S. 20-25
6. Lazzereini B., Maggiore A., Marcelloni F., Classification of odour samples from a multisensorarray using new linguistic fuzzy method, Electronic letters, Vol. 34, No. 23 pp 2229-2231 (1998)

Section 5: Applications: Food, Agricultural and Environmental
Paper presented at the Seventh International Symposium on Olfaction and Electronic Noses, July 2000

279

Use of Chemical Sensor Arrays for Recognition of Odour Emissions from Agriculture

Gisbert Riess, Barbara Maier

*Bayerische Landesanstalt für Landtechnik, TU München, 85350 Freising, Germany
Phone: ++49 8161 715341, Fax: ++49 8161 714363; email: riess@tec.agrar.tu-muenchen.de*

Abstract: It was investigated whether a multisensor array is suitable to measure and classify odour emissions from farms. It was demonstrated that the multisensor array is able to distinguish the odour emissions from agriculture. Some examples are shown in this article. It was possible to separate odours samples of different animal species. The multisensor array is also able to distinguish the odour emissions of different housing systems for the same animal species like fattening pig husbandry with different ventilation systems. A distinction of the odours was made between dairy cattle and beef bulls. In this case, a stall air sample that was taken at a later point was also reliably classified in the right class. It was shown that a neural network is most suitable for such a valuation.

1. Introduction

Due to the increasing density of the population and the growing demands with regard to air quality, the registration of odour nuisance has become a problem that may no longer be neglected. Even if the concentrations of odour-causing substances are generally so low that they do not pose any danger to health, they often give rise to considerable annoyance. Since authorities must deal with an increasing number of complaints in permission procedures, the development of a system for continuous odour measurement is necessary. Therefore new systems based on multisensor arrays were developed to measure odour emission from agriculture [1-4].

2. Experimental

A multisensor array from Alphamos (FOX 4000) with 18 semiconductor sensors was used. The odour samples were taken into Tedlar bags with a size of about 30 litres. Afterwards the measurements were made in the laboratory [3,4]. For use outside the laboratory, the whole sensor system including the humidification device was put in a heat insulated box. This outdoor system was located on a central point of the trial farm.

3. Results

3.1 Recognition of Odour Emissions from Stalls with Different Animal Species

Samples were taken from a piggery, a cattle-, and a chicken stall and analysed by the multisensor array. The measurement values are shown as PCA-plot (figure 1). The goal is to achieve a clear separation of the three groups. The multisensor array is able to separate the odour emissions of a piggery from the odour emissions of a cattle and a chicken stall. The human nose can also make this distinction, i.e. in the case of this application, the multisensor array achieves the same result as the human nose.

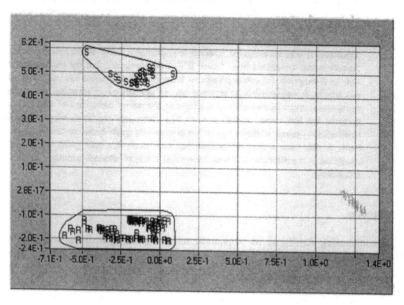

Figure 1: PCA-plot based on samples from a pig-(S), cattle-(R), and chicken (H) stall

3.2. Comparison of Ventilation Systems for Fattening Pig Husbandry

The experiment was done on a trial farm. The multisensor array was connected with the ventilation system for about 10 days. The building of one trial farm that was examined contained two compartments with 200 places each for fattening pigs. The animals are fed mash, and the stall is mucked out according to the damming-washing method. In one compartment, the outgoing air was sucked off by underfloor suction (U-flur) through a longitudinal channel under a central walkway. Fresh air is supplied by 5 central ceiling distributors in the longitudinal axis of the respective stall compartment. In the other compartment, the ventilation system was changed to overfloor suction (O-flur) by placing the suction point above the level of the slatted floor. Figure 2 illustrates the trial set-up.

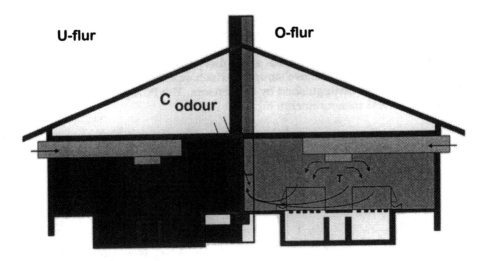

Figure 2: Diagram of the two ventilation systems compared in fattening pig husbandry

The results of the measurements of the multisensor array show the qualitative difference of the odour emissions from both stall compartments. As shown in figure 3, the gas samples from both stall compartments with their different ventilation systems are separated. In this case the multisensor array discriminate the odour emissions between the ventilation system.

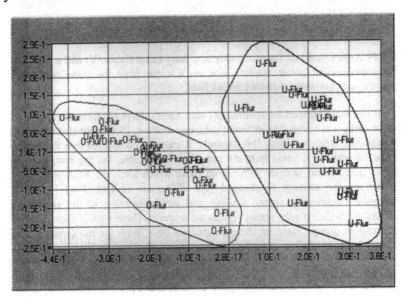

Figure 3: PCA-plot of the samples from different Ventilation Systems for Fattening Pig Husbandry

3.3. Qualitative Results: Distinctions with the Multisensor Array between Different Cattle Stables / Comparison of Different Valuation Methods

Samples were taken in a beef bull- and a dairy cattle stable and analyzed by the multisensor array. The two samples, which smelled very similar to the human nose, were distinguished by the sensors. The PCA-plot shown below is based on these measurements (figure 4).

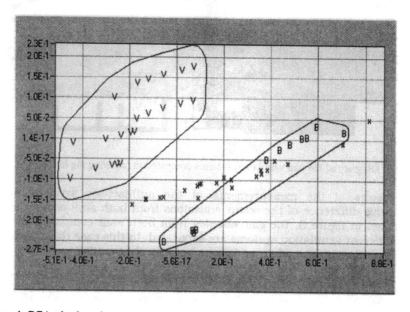

Figure 4: PCA-plot based on measurements in a dairy cattle stable (V) and a beef bull stable (B). The measurement points marked with an X come from the same beef bull stable. However, the samples were taken four weeks later

On the basis of these samples from different cattle stables the possibility of recognition by the multisensor array was tested. The different valuation methods [5,6] were also compared based on this example. When assessing this example, it must be considered that the measurement points marked with an X come from the same beef bull stable, but the samples were taken four weeks later. These samples differ from those which form the PCA-plot, because the composition of the samples is changed. The difference is caused by factors such as meteorological fluctuations, altered feeding, and weight increase of the animals. For the unmonitored methods, only the sensor signals are relevant. These signals classify the different measurements in groups based on the measurement data. With the monitored methods, the groups must be known before valuation. During the actual valuation, the measurement values are assigned to the groups. Of the above described methods, those that were selected for data valuation are included in the FOX 4000 software package used. The represented Plot (figure 4) shows the result of principal component analysis (PCA), an unmonitored, model-based valuation method. Here, 70% of the measurements of the unknown sample (X) are assigned to the right group (beef bull stable B). Valuation with discriminating factor analysis (DFA), a monitored, model-based method, enabled 90% of the newer measurements to be matched to the corresponding measurements that were four weeks older. The highest recognition rate of

95% was achieved by a trained neural network with backpropagation architecture. The results of the comparison of the valuation methods are again summarized in table 1.

Table 1: Comparison of valuation methods using different cattle stables as examples

valuation method	recognition rate in percent
principal component analysis (PCA) (unmonitored, model-based)	70 %
discriminating factor analysis (DFA) (monitored, model-based)	90 %
neural networks with backpropagation architecture (BPN) (monitored, model-free)	95 %

The table demonstrates that the samples which were taken four weeks later were recognized. The neural network leads to highest recognition rates. Therefore this neural network is most suitable for such a recognition.

4. Conclusion

The results documented that odours from agriculture can be distinguished by the multisensor array. It was possible to separate odours of different animal species. The multisensor array is also able to distinguish the odour emissions of different housing system for the same animal species like fattening pig husbandry with different ventilation systems. In addition, a distinction was made between dairy cattle and beef bulls In this case, a stall air sample that was taken at a later point was also reliably classified in the right group. It was shown that a neural network is most suitable for such a valuation. The big advantage of this system is that it monitors odours continuously in relatively small intervals. This opens up the possibility to observe the development over the course of the day in short intervals and at the same time offers the chance to monitor odours continually over a period of several weeks. Therefore it is possible to asses different agricultural production processes according to the odour emissions. The production system which leads to less annoyance can be chosen.

5. Acknowledgement

This work has been supported by the Bayerischen Staatsministerium für Landesentwicklung und Umweltfragen (BayStMLUF).

6. References

1. Schiffman S.S., Classen J.J., Kermani B.G.,Troy Nagle H.;International Conference on Air Pollution from Agricultural Operations, Missouri, S. 255-261 (1996)
2. Hahne J., Vorlop K-D., Hübner R., Müller:D; Landtechnik 54-6, S. 350-351 (1999)

3. Maier B., Rieß G., Gronauer A.; Landtechnik 55-1; S. 44-45 (2000)
4. Riess G., Maier B., Gronauer A.; Landtechnik 55-2, S. 176-177 (2000)
5. Dillon W.R., Goldstein M.: Multivariate Analysis Methods and Applications; John Wiley and Sons, Inc New York (1984)
6. Kraus G., Weimar U., Gauglitz G., Göpel W.; Technisches Messen Vol.62, S.229-236 (1995)

Using a Portable Electronic Nose for Identification of Odorous Industrial Chemicals

C Furlong [1] and J R Stewart [2]

[1]QUESTOR Centre and [2]School of Computer Science, Queen's University of Belfast, Belfast, BT7 1NN, Northern Ireland.
Phone: +44 (0)28 9033 5405, Fax: +44 (0) 28 9058 3890
E-mail: c.furlong@qub.ac.uk

Abstract. This paper describes some initial experiences in using a prototype portable electronic nose for the detection and recognition of odorous materials. Principal component analysis techniques were successfully used to demonstrate a high identification rate for some common industrial chemicals in simple vial tests without the need for environmental controls.

1. Introduction

The aim of this project is to determine the value of using an electronic nose for the identification of fugitive odours arising from industrial activities. A prototype of a potentially suitable portable electronic nose (illustrated in Figure 1) has been provided by Cyrano Sciences Inc.

The nose has a detector array containing thirty-two carbon black/polymer composite sensors [1-4], four temperature sensors and a relative humidity sensor. The carbon black forms the conducting phase of the sensor, which is dispersed into an insulating organic polymer. When the polymer comes into contact with an organic vapour it swells causing a change in the electrical resistance of the sensor. An electrical potential is applied across each sensor so that the resistances may be recorded.

Figure 1: Prototype electronic nose developed by Cyrano Sciences Inc.

The use of different polymers gives the sensors a range of sensitivities to a variety of chemicals. It has been reported [5] that the sensors can show a response to some compounds at concentrations as low as 0.5% of their saturated vapour pressure and that a mixture of compounds will exhibit an additive response.

The commercial version of the nose will be a small hand-held battery-powered model. The pattern recognition system software supplied with the prototype nose, known as *distance vectors*, operates by finding the minimum difference between the magnitudes of the responses to an unknown and those for already known compounds. Responses to known materials are stored and then used for the identification of unknown samples. The commercial version will use a more sophisticated pattern recognition system based on principal component analysis (PCA).

2. Experimental Procedure

Small vials (25 ml) were filled with 5 ml of the compounds listed in Table 1. The nose was first 'trained' by sampling each compound once and storing the response patterns. Ten samples were then taken for each compound alternating between compounds within each experimental group. The tests were carried out without using temperature or humidity control, laboratory air was taken for a baseline and the concentration in the headspace of the vial was unregulated. All results were analysised using the statistical package SPSS© 7.5.

Table 1:The compounds within each experiment group with their vapour pressures, saturated concentrations and odour detection thresholds.

Experiment Number	Compound	Vapour Pressure (mm Hg)	Saturated concentration (ppth)	Odour detection threshold (ppth)[6]
Experiment 1	toluene	26	33	0.16E-3
	isopropyl alcohol (IPA)	44	55	0.442E-3
Experiment 2	acetone	400	345	4.58E-3
	methyl ethyl ketone (MEK)	71	85	0.27E-3
	2-butoxyethanol	1	1	63E-6
	cyclohexane	169	182	83.8E-3
	ethyl acetate	73	88	0.61E-3
	ethanol	100	116	0.136E-3
	methanol	410	350	4.05E-3 - 75E-3
Experiment 3	caproic acid	0.18	0.2	3.86E-6
	valeric acid	0.15	0.2	1.8E-6 - 2.63E-3
	butyric acid	0.43	0.6	89.09E-9 - 21.9E-6
	acetic acid	11.4	15	16E-6
Experiment 4	ammonia (32% solution)	115	131	-
	acetic anhydride	10	13	0.29E-6
	o-xylene	7	9	162E-6 - 4.89E-3
	acetonitrile	72.8	87	-
	tetrahydrofuran (THF)	134	150	-
	acrolein	817	518	1.4E-9
Experiment 5	formic acid	44.8	56	976E-6
	propionic acid	2.9	4	1.51E-6 - 176E-6

3. Results

The response for each of the thirty-two sensors to a sample was taken to be the change in resistance of the sensor divided by the baseline resistance recorded with background air. The distance vectors used to identify the compounds were calculated by the distance between the response of the array to a sample and the response of the array to each of the target compounds.

Ninety-four percent of the samples were identified correctly using this method. Half of the samples of acetic anhydride and MEK were misidentified.

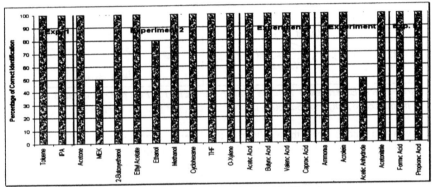

Chart 1: Percentage of Correctly Identified Samples.

4. Discussion

The experiments were carried out over a nine month period during which time several of the sensors failed. Any results obtained from these sensors were omitted from the analysis presented here if they were not working fully throughout an experiment. It is suspected that this failure was an electrical fault, as all of the failed sensors were on the same bank and were connected from the same supply, but it may also have been that the sensors were 'poisoned' by the high concentrations of compounds used.

Overall, the prototype nose performed extremely well. The high success rate may have been facilitated by the high concentrations of the compounds in the sample headspace. It was assumed that the headspace of the vial was saturated and the concentrations varied from approximately 0.2 parts per thousand (ppth) for caproic acid to 518 ppth for acrolein (see Table 1). These concentrations are high in comparison to odour detection thresholds. It is apparent that the statistical method used (distance vectors) in the prototype electronic nose was usually adequate for identification of the samples at these relatively high concentrations.

The results of Experiment 4, illustrated in Chart 2, show an 'outlier' among the responses for acetic anhydride. The outlier, which is plainly not representative, was the response stored when the nose was trained on this compound and this explains the poor performance in identifying acetic anhydride. When a more appropriate standard was set all of the subsequent samples were correctly identified. It is not known why this first standard was unrepresentative. It could have been due to uncontrolled conditions when sampling, but it has also been noted that the sensors need to be used (i.e. flushed by air or exposed to a sample) before a representative response is given. The omission of this pre-conditioning may cause problems if only one target sample is used.

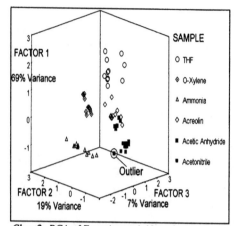

Chart2: PCA of Experiment 4. Note the 'outlier' in the Acetic Anhydride Sample set.

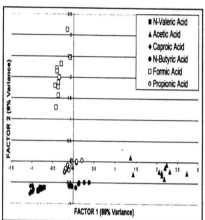

Chart 3: PCA of Experiments 3 & 5 Combined.

PCA was also used to assess the results from the different experiments, as in Chart 3, where all of the organic acids are separable in PC space. This suggests that, if a further single experiment was to be carried out using all of the organic acids in experiments 3 and 5, it would be expected that all samples would be correctly identified.

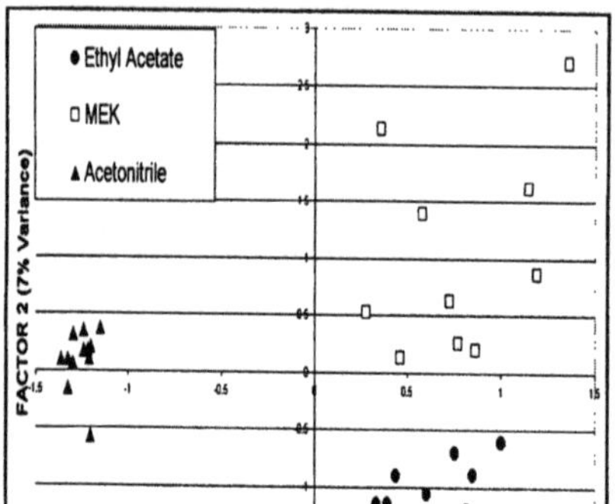

Results from the test compounds with similar saturated vapour pressures of approximately 70 mm Hg (acetonitrile, ethyl acetate, and MEK) were analysed using PCA. The vial method means that these compounds would have been sampled at very similar concentrations. From the results of this analysis in Chart 4 it can be seen that the compounds are easily separable in PC space. This illustrates the selectivity of the array of sensors. The sensors responses are not solely related to the concentrations of the compounds, they are responding selectively to the sample compounds. This can also be seen in Chart 3 as caproic and valeric acids have similar saturated vapour pressures and are easily separable in PC space.

Chart 4: PCA of Compounds with similar vapour pressures of about 70 mm Hg.

PCA of the results from experiment 2, illustrated in Chart 5, show that MEK overlapped the PC space of acetone. The reasons why acetone was not identified as MEK are not obvious as the MEK results had a lower total standard deviation than those for acetone, (MEK 1.59 and acetone 1.82). When both fingerprints are normalised to 1 (by dividing the response of each sensor by the total response of all sensors for one exposure), it can be seen that they are very similar, see Chart 6. It had been anticipated that MEK would have a higher total standard deviation, as it was identified as acetone, but acetone was not identified as MEK. When the results were normalised, removing the bias due to differing concentrations, the order of the total standard deviations was reversed with MEK having a slightly higher total standard deviation of 0.0962 than acetone with 0.0782. This reflects the results reported. The similarity in the fingerprints is undoubtedly due to the similarity in the chemical structure of the two compounds acetone – CH_3COCH_3 and MEK-$CH_3CH_2COCH_3$, even though some of the physical properties of the compounds are different.

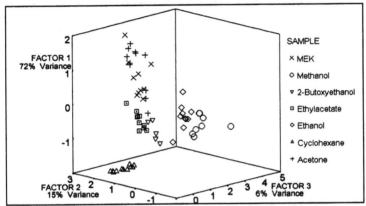

Chart 5: PCA of Experiment 2. Note the results of MEK and Acetone

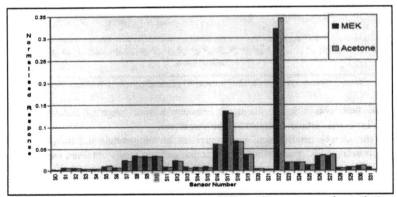

Chart 6: The Normalised Average fingerprints of MEK and Acetone: note the similarity.

The drift of the array response over time is being monitored using cyclohexane as an indicator compound. Cyclohexane was chosen as the arrays' response was very reproducible, this can be seen in Chart 5 as the cyclohexane samples are closely clustered in PC space reflecting only a slight variance in the sensor response to the samples. The drift over a nine-month period can be seen in Chart 7. (The exact age of the sensors is not known.)

The percentage drift of the sensors was calculated from the change in the response of each sensor using the first and last results of the nine-month period. This gives an overview of the drift pattern. It can be seen that there has been little drift over this time period. The averaged drift of the array was +9%. The drift can be either positive or negative, sixteen sensors drifted negatively (their response decreased over time) and eleven sensors drifted positively (their response increased over time). The characteristics of the drift may be linked to the polymer used in the sensor. The only sensor that drifted more than 100% was S31 (+340%), this extreme drift could indicate sensor failure.

5. Conclusion

The prototype electronic nose shows a high success rate in differentiating between the samples tested for this work. This may be partly attributed to the high concentrations used. No environmental control of either temperature or relative humidity was needed to achieve these excellent results and this represents a useful advantage over other types of sensor. The statistical analysis carried out by the software provided with the prototype electronic nose was sufficient for the samples to be identified. PCA proved to be a valuable tool for analysing the results in more detail and the new commercial version of the nose will use this and other sophisticated statistical methods.

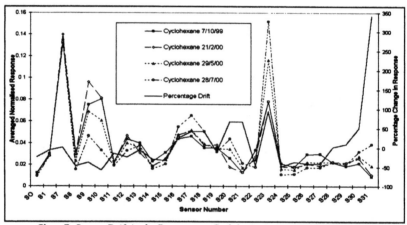

Chart 7: Sensor Drift in the Response to Cyclohexane over a 9-Month Period.

The nose was able to distingush between a wide range of industrial chemicals including those with similar saturated vapour pressures.

If the sample environment was controlled for temperature and humidity a smaller spread of results in PC space would be expected and the results obtained would show less variability. This might avoid the low identification rates for MEK.

References

[1]   Lonergan M C, Severin E J, Doleman B J, Beaber S A, Grubbs R H & Lewis N S 1996 *Chemistry of Materials* **8** 2298-2312.

[2]   Doleman B J, Sanner R D, Severin E J, Grubbs R H & Lewis N S 1998 *Analytical Chemistry* **70** 2560-2564.

[3]   Severin E J, Sanner R, Doleman B J & Lewis N S 1998 *Analytical Chemistry* **70** 1140-1143.

[4]   Doleman B J, Lonergan M C, Severin E J, Vaid T P & Lewis N S 1998 *Analytical Chemistry* **70** 4177-4190.

[5]   Severin E J, Doleman B J & Lewis N S 2000 *Analytical Chemistry* **72** 658-668.

[6]   Woodfield M and Hall D 1994 Odour measurement and control-An update AEA Technology, Oxfordshire.

Section 5: Applications: Food, Agricultural and Environmental
Paper presented at the Seventh International Symposium on Olfaction and Electronic Noses, July 2000

291

Monitoring of Hot Flue Gases by an E-Nose Equipped with SiC Based Sensors and Metal Oxide Sensors

L. Unéus[*], M Mattsson[*], P. Ljung[*], R. Wigren[**], P. Mårtensson[**], L. G. Ekedahl, I. Lundström, A. Lloyd Spetz.

S-SENCE and Division of Applied Physics, Linköping University SE-581 83, Linköping, Sweden.
[*]Vattenfall Utveckling AB, SE-814 26, Älvkarleby, Sweden.
[**]Nordic Sensor Technologies AB, Teknikringen 6, SE-583 30 Linköping, Sweden
Phone: +46 13 288904 Fax: +46 13 288969 E-mail: larun@ifm.liu.se, http://www.ifm.liu.se/Applphys/S-SENCE/

Abstract. We have used Metal Insulator Silicon Carbide (MISiC) sensors, Metal Oxide Sensors (MOS) and a linear lambda sensor in an electronic nose to measure hot flue gases on-line. Several reference instruments, which measured the flue gases in parallel to the sensor array, were connected to the electronic nose. Flue gas from a 500 kW pellets fuelled boiler, which is used for heating apartment blocks, has been used to feed the experimental set-up. The gases which are interesting to measure is NO, CO, O_2 and hydrocarbons (HC). On-line quantification of these gases would make it possible to regulate the boiler for a better power economy, which would also reduce the emissions of hazardous gases. Results on prediction for CO, NO, O_2 in the flue gases based on PLS models are presented.

1. Introduction

There is a great need from industry today for a cheap way of monitoring the amount of gases in the exhaust from oil burners in small power plants since legislation's are getting stricter in the near future. Equipment like FTIR instruments used in larger power plants are too expensive for boilers of 0.5 – 5 MW. However, there are other possible solutions to the problem [1] Here we have used a specially built electronic nose, which consists of 12 MISiC sensors, 16 MOS and one linear lambda sensor [2]. Also connected to the electronic nose data acquisition program, are several reference instruments, measuring CO, CO_2, NO, O_2 and HC, and three parameters from the boiler regulation system; speed of fuel, combustion temperature and the temperature of the outgoing hot-water.

Using multivariate methods we correlate different parameters (sensor signals and boiler signals) to the different gas concentrations in the flue gases. By building PLS (Partial Least Square) models, predictions of some of the gas components (CO, NO, O_2) in the flue gases can be made.

2. Experimental

2.1 *Measurement equipment*

The MiSiC sensor consists of a field effect Schottky diode with gate contacts of Pt or Ir. These metals are deposited on SiC that has been ozone cleaned before the metal deposition [3]. The Silicon Carbide (6H SiC) wafers are n-type (nitrogen doped to 3.3×10^{18} cm^{-3}) with a 10 µm n-type epilayer (2.6×10^{16} cm^{-3}) [4], and ohmic back contacts of 200 nm TaSi$_x$ + 400 nm Pt. The precise processing is described elsewhere [5]. The MOS and the linear lambda sensor are commercially sensors [6,7,8], and the linear lambda sensor is specified for oxygen concentrations between 0 and 12.5%.

The special high temperature electronic nose, which has been constructed for this project [2] has a "sensor box" that is connected to the electronic part of the nose. The sensor box contains two sensor blocks, one with 12 MiSiC sensors and one with 16 MOS. The sensor blocks are heated and regulated at 200°C to prevent the dirty and sticky flue gases to build up contaminating soot-layers inside the blocks. The flue gases are transported to the sensor box through heated tubes, are also kept at 200°C. Using heated blocks are cumbersome; the high temperature is causing a lot of problems since electronics seldom are specified to be used over 100°C. This have forced us to make several special solutions, which have led to a sturdy design and easy to use sensor box, which can easily be connected to any gas flow. With the name "high temperature electronic nose" we refer to the fact that we are using hot gases and that the nose is designed to handle that, but the sensors are also operated at high temperature (higher then the hot flue gases).

The reference instruments connected to the electronic nose are; one paramagnetic cell working as an oxygen sensor, a combined CO and CO$_2$ IR-instrument, which can measure from 0 to 500 ppm CO, a NO instrument and a Flame Ionisation Detector (FID). The experimental set-up can be seen in Fig. 1.

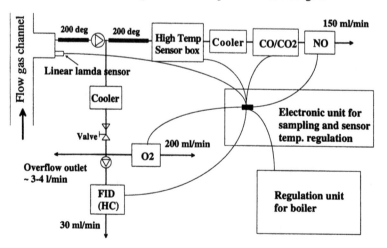

Figure 1. The experimental set-up, including the part of the flow gas channel where the gases are pumped out and into the measurement equipment.

2.2 *The MISiC Sensor principle*

When MISiC Schottky diode sensors are exposed to hydrogen, hydrogen decomposes on the surface and the hydrogen atoms diffuse through the metal and form a polarised layer at the metal-oxide interface. This polarised layer shifts the current voltage, IV, curve to a lower voltage, and the gas response is measured as a shift in the forward bias at a constant current [9]. The sensing mechanism for porous metals is more complicated. The main theory is that the molecules adsorb on the metal and react with oxygen. Reaction products diffuse out on to the oxide in the holes and cracks of the metal film. Molecules or reaction products with strong dipoles should be able to change the shift in the forward bias [5].

3. Results

In results from earlier measurements we found some MISiC sensors with good sensitivity and selectivity towards CO. Also O_2 was predicted rather well in the flue gases, mostly due to the linear lambda sensor [10].

 The intention of the new and improved electronic nose, was to make predictions for CO, O_2, NO and HC. It turned out that the HC level in the flue gases are very low, only a few ppm, which is out of range for our electronic nose. The measurements were now directed to prediction of CO, NO and O_2 using PLS models.

4.1 *The flue gases*

Since the boiler is continuously heating apartment blocks, there is a natural randomisation of the flue gases. Since it is chemical combustion, there is of course a correlation between the gas components. However, the correlation is not large enough to be used for regulation of the boiler by only one gas component, i.e. several gas components must be measured. In Figure 2 typical CO, NO and O_2 concentrations are seen. The large relative variation for CO makes it easier to predict, however, at the same time it makes it harder to measure NO and O_2, since the large CO variation "hides" the smaller variations for NO and O_2.

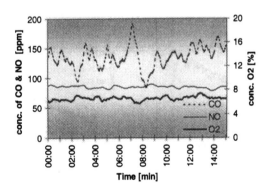

Figure 2. Typical gas concentrations in the flue gases over a 15 min period. The relative change in CO is much larger than the variation of NO and O_2.

4.2 CO measurements and predictions

The variation of CO in the flue gases can be seen in Fig. 2. The mean value is typically around 150 ppm with the maximum around 500 ppm and the minimum around 80 ppm. This means that the variation is very advantageous for measuring and prediction of CO since $\Delta CO/CO_{max}$ is very high.

The electronic nose is running round the clock and the sampling is made every 4:th second. This means that there is a lot of data and everything cannot be used for evaluation. Three hours of data (2700 samples) from one day have been used for the evaluation. With these 2700 samples a PLS model for CO has been built with two third as calibration data and one third as validation data. The model is then used for prediction of CO, for example one day later, one week later and so on.

A PLS model with 2 principal components (PC's) was built, with 2 MISiC sensors, 3 MOS, the linear lambda sensor and the temperature in the oven as the X-variables and the CO concentration from the IR-instrument as the Y-variable. No special pre-treatment have been done on the dataset, except for auto-scaling, i.e. mean centred and divided by the standard deviation. The Root Mean Square Error value (RMSE-value) for the validation data was around 7.5, which means that predictions on validation data can be made with an accuracy of approximately ± 15 ppm. However, predictions for future measurements are most interesting. We have used the model to predict concentration the next day and after five days and compared to measured data. The results can be seen in Fig. 3 after one day and in Fig. 4 after five days as predicted vs. measured graphs.

The R^2-values for both the graphs are quite good, which indicates that the model is valid up to one week. The offset is very small in both cases, which is good. However, the slope in Fig. 4 deviates from 1. This is due to the fact that the model cannot predict high concentrations of CO, both because the model is based on a data set without the extreme high CO concentrations, and that the sensor response is non-linear in that region.

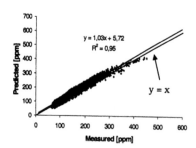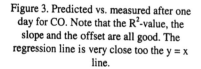

Figure 3. Predicted vs. measured after one day for CO. Note that the R^2-value, the slope and the offset are all good. The regression line is very close too the y = x line.

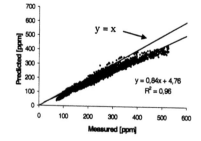

Figure 4. Predicted CO vs. measured after five days for CO. Note that the R^2-value and the offset are as good as after one day but the slope has change.

4.3 *NO measurements and predictions*

To measure NO in flue gases or in lean burn engines are of great interests not only in this paper [11]. The variation of NO in the flue gases is seen in Fig. 2. The mean value is typically around 80 ppm, the maximum is around 100 ppm and the minimum is around 60 ppm. This means that the variation is not very advantageous for measuring and prediction of NO since $\Delta NO/NO_{max}$ is very low compared to the ratio for CO.

Since the sensors are more or less sensitive towards all three relevant gas components in the flue gases (CO, NO and O_2), most of the sensor responses originate from the variation in the CO concentration, and the NO and the O_2 concentrations in the flue gases are more difficult to measure. However, some promising results have been obtained for NO, which indicates the possibility to predict NO in the flue gases.

A PLS model with 3 PC's has been made, which has 4 MISiC sensors, 4 MOS, the linear lambda sensor and the fuel speed as the X-variables, and the NO concentration from the IR-instrument as the Y-variable. No special pre-treatment was made on this dataset, except for auto-scaling.

The predicted vs. measured plots after one day and after five days can be seen in Fig. 5 and 6 respectively. After one day the values are acceptable, but after five days the model has some problems to predict the right value. This can be due to several factors: the sensors have some drift, or the gas composition has changed so the model does not recognise the mixture any longer. If the problem is drifting sensor it might be solved with drift correction methods, and if the problem is the flue gas composition, a better and more robust model must be used. To achieve a better model, a larger and more representative data set must be used for building the model.

It should also be noted that a more optimised operating temperature on the MISiC sensors could also improve the results. At this point in the measurements, not all possible temperature combinations have been tested. However, the results are still promising considering these circumstances.

4.4 *O_2 measurements*

Due to malfunction on a pump in the reference oxygen equipment, very few good measurements have been performed. However, it was possible to make a PLS model for O_2 prediction but no predictions on data have been done. The same problem as with the NO measurements applies for the prediction of the O_2 concentration, i.e. it is difficult to see the small change in the O_2 concentration compared to the large CO variation.

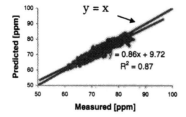 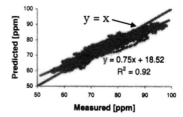

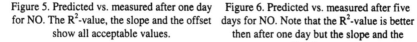

Figure 5. Predicted vs. measured after one day for NO. The R^2-value, the slope and the offset show all acceptable values.

Figure 6. Predicted vs. measured after five days for NO. Note that the R^2-value is better then after one day but the slope and the offset is worse.

The preliminary PLS model for O_2 has 4 PC's, 3 MISiC sensors, 2 MOS, the linear lambda sensor, the fuel speed and the temperature in the oven. The RMSE-value for the validation data is 0.23, which means that the O_2 concentration can be predicted with $\pm$ 0.5%. The R^2-value is 0.87 for the validation data, the slope is 0.81 and the offset is 1.35, which means that the model needs improvement. However, considering that very few measurements have been done, no optimisation of the operating temperature for the MISiC sensors is made and no pre-treatment of the data is made (only auto-scaling), the result are quite promising and the O_2 prediction will probably be successful in the future.

5. Summary

We have demonstrated an electronic nose successfully operating in hot flue gases. We have shown that CO can be predicted with very good precision in the flue gases for at least five days using the same simple PLS model. NO and O_2 show very promising results considering such very small variation of these gases compared to the variation in the CO concentration.

More measurements will be performed and including more detailed studies on the optimisation of the operating temperature for the MISiC sensor. Also drift correction methods will have been tested on the long-term measurements.

6. Acknowledgements

Our research on high temperature chemical sensors based on silicon carbide is supported by grants from the Swedish National Board for Industrial and Technical Development and Swedish Industry through the centre of excellence, Swedish Sensor Centre, S-SENCE. This work was carried out at Vattenfall Utveckling AB in Älvkarleby, Sweden and financed by Vattenfall Generation Service AB. Nordic Sensor Technologies built the high temperature electronic nose.

References

[1]   Hargreaves KJA and Maskell WC
[2]   Nordic Sensor Technologies AB, Teknikringen 6, Mjärdevi Science Park, SE-583 30, Linköping, Sweden
[3]   Cree, Research Inc. 4600 Silicon Drive, Durham, NC 27703, USA
[4]   Unéus L, Tobias P, Salomonsson P, Lundström I, Lloyd Spetz A 1999 Sensors and Materials 11 5 305-318
[5]   FIGARO ENGINEERING INC, 1-5-11, Sensbanishi, Mino, Osaka, 562, Japan
[6]   FiS INC, 2-5-26, Hachizuka, Ikeda, Osaka, 563, Japan
[7]   Robert Bosch GmbH, Postfach 10 60 50, D-70049, Stuttgart
[8]   Lundström I, Shivamaran S, Svensson C, Lundqvist L 1975 Appl. Phys. Lett. 26 55 3876-3881
[9]   Unéus L. Ljung P. Mattsson M. Mårtensson P. Wigren R. Tobias P. Lundström I. Ekedahl L. G. and Lloyd Spetz A. Proc. Eurosensors XIII, (1999), The Hague, The Netherlands, 26A3
[10]   Ménil F. Coillard V. and Lucat C 2000 Sensors and Actuators B 67 1-23

Quartz Crystal Microbalance Sensors Based Electronic Nose for QC in Automotive Industry

S. Garrigues[1], T. Talou[1] and D. Nesa[2]

[1] Laboratoire de Chimie Agro – Industrielle, ENSCT, 118 rte de Narbonne, 31077 Toulouse cedex 4, France
[2] Technocentre RENAULT, 1 av. du Golf, 78288 Guyancourt cedex, France
Phone: +33562885738 Fax: +33562885730 E-mail: sgarrigues@ensct.fr

Abstract. The automotive industry continuously tries to satisfy customers expectations in comfort requirements. To traditional requests like visual or tactile aspects, a new dimension appear to be an important criteria for customer's choice: smell. Due to their manufacturing process and to their petrochemical compounds based, many rubbers and foams used in automotive materials result in a "new car odor" mostly enjoyed but sometimes felt as unpleasant by customers. This paper presents the different results obtained on these interior trim materials using a quartz crystal microbalance sensors based electronic nose (QMB 6 device) after optimization of static headspace generation performance by using experimental design methodology.

1. Introduction

Since a few years, the car industry which is confronted with a lot of parts that contribute to the "new car odor" [1] has become a new field of application of electronic nose technology. Most of these parts included polymers based elements like polyurethane foams for the seats, polypropylene and PVC for the dash boards and the doors... Parallely a lot of glues are used to seal the parts together and organic coatings which preserve the body. Therefore, the automotive industry strongly need to control easily and rapidly those different parts in order to decrease the number of potential sources of unpleasant odors and to create a more pleasant and safe environment for the users. Different types of gas sensors have been already tested on this kind of materials (MOS, MOSFET and CP sensors) [2, 3, 4] and the measured responses were based on electrical resistance shifts.

Acoustic-wave-based microsensors included 2 kind of sensors : SAW (Surface Acoustic Wave) and BAW (Bulk Acoustic Wave) also called QMB (Quartz crystal MicroBalance). Quartz crystal microbalance is commonly configured with electrodes on both sides of a thin disk of AT-cut quartz in the opposite of SAW constituted of 2 planar electrodes. Due to the piezoelectric properties and cristalline orientation of the quartz, the application of a voltage between these electrodes results in a shear deformation of the crystal [5]. Acoustic-wave-based microsensors obtain their chemical sensitivity and selectivity from a chemically active coating, which interacts with the surrounding environment. Analyte sorption in the coating is considered as pure mass accumulation and this interaction leads to change in the acoustic properties at the interface between the acoustic device and the coating, which in turn, yields a change in the electrical response of the sensor i.e shift of the resonant frequency [6, 7]. Quartz crystal microbalance sensors based electronic nose is the chosen technology to analyse and discriminate these different polymer materials from car interior.

2. Experimental Procedure

2.1. *Apparatus*

The QMB 6 sensor system consists of a chemosensor array based on quartz crystals (HKR Sensorsysteme, München). 6 sensor elements are integrated on a common quartz substrate. These sensors are coated with different gas-sensitive materials (polar and apolar GC coatings) that react differently to the gases to be analysed and are built into a temperature-controlled measuring chamber. Each oscillator has a basic frequency in the range of 10 MHz and the resolution of the sensor signal is ± 1 Hz.

The sensor signals were processed by multivariate algorithms. Principal Component Analysis (PCA) was used for classification of the samples (QMB 6 Sensorsysteme software).

2.2. *Feasability study*

10 various polymer samples from car interior were analysed (15 replicates) using the above technology (foams, interior trim materials, dash board). Samples were packaged in individual specific plastic bags filled under nitrogen and refrigerated at +4°C during storage [8].

Static headspace sampling was carried out with the HS 40 XL autosampler (Perkin Elmer, Norwalk). Polymer samples (12 mm circles) are introduced in 22 ml vials and heated up during a certain time. After this period of time, an equilibrium of headspace is established : stable partial pressures of the different gases over the sample are reached. The vial is pressurized (1 min) using an inert gas, and 1 ml of the headspace vapors released into the measuring chamber, containing the sensors, through a capillary tube.

2.3. *Headspace optimisation*

Experimental design methodology is used in order to optimise static headspace analysis performance.

The studied sampling parameters are : temperature [30°C - 131°C], equilibrium time [120 min - 960 min], ratio sample volume/gas volume [1/15 - 7/15], and recorded responses are at one and the same time peaks area and number of peaks. A central composite design was set up (20 experiments : 2^3 + 6 star points + 6 replicates at the centre) and data were processed both by NEMROD (LPRAI, Marseille) and UNSCRAMBLER (CAMO ASA, Oslo) allowing the drawing of the isoresponses curves.

2.4. Analysis of 7 different PVC batches (after headspace optimisation)

Measurements (6 replicates) of 7 different PVC batches issued from various suppliers and manufacturing processes have been done. Experiments are performed under optimised conditions.

To avoid insidious effects may be produced by the position in order of presentation of samples and by carry-over from one sample to the next, randomisation of the vials is carried out. Failure to take account of these effects can lead to biased results, and consequently to misinterpretation of the results of analysis of variance, preference mapping or principal components analysis [9].

3. Results and Discussion

3.1. Feasability study

As it is obvious from the graph, quartz crystal microbalance sensors are able to discriminate between some of the polymer samples more distinctly than other (e.g. PVC and AR-PV which are fibrous constituted samples).

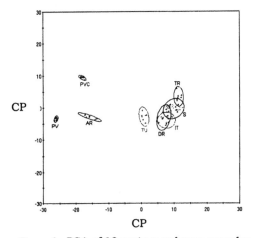

Figure 1 : PCA of 10 various polymer samples

3.2. Headspace optimisation

Figure 2 shows that as expected, temperature seems to be the most significant parameter. However, a technical constraint concerning exposure temperature of the different automotive trim materials was set by industrialists at 80°C. Isoresponse curves have allowed to determine the optima of the main parameters governing the headspace generation in the studied experimental field. They are listed in table 1.

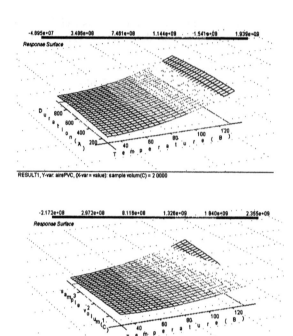

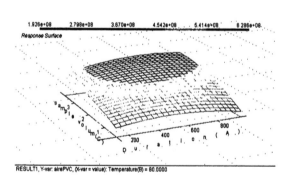

Figure 2 : Response surfaces from PVC experimental design

Table 1 : Optima conditions obtained in the experimental field

Duration	A	540 min.
Temperature	B	80°C
Sample Volume	C	1/3 of the vial

3.3. *Analysis of 7 different PVC batches (after headspace optimisation)*

As it shows in Figure 3, quartz crystal microbalance gas sensors allow to discriminate between different PVC batches .

The obtained differences using electronic nose corroborate the observed ones by a global sensory analysis (tactile, visual and olfactive).

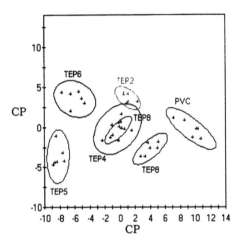

Figure 3 : PCA of 7 kind of PVC samples

4. Conclusions and outlook

The work presented in this investigation shows that QMB gas sensors seems to be sufficiently sensitive to discriminate under optimised headspace generation conditions between different polymer materials. The experimental design used has permitted to improve QMB 6 sensors system performances by optimising headspace generation.

These preliminary results were very promising but must be related with gas chromatography and sensory analysis. A trained panel is presently performing sensory analysis using "The Field of Odors" olfactory referential [10], and additional measurements with QMB are actually underway in order to validate another statistical treatment tool, such as ANNs [11].

References

[1] K.K. Sakakibara, C. Hamada, S. Sato, M. Matsuo, *Analysis of odor in car cabin.* JSAE 9930676 **(1999)** pp. 237-241.
[2] E. L. Kalman, F. Winquist and I. Lundström, *Gas emissions from car interior trim materials measured by an electronic nose,* 6th ISOEN Proceedings, Tübingen (Germany) **(1999)** pp. 384-386.
[3] M. Morvan and T. Talou, *Electronic nose systems for control quality applications in automotive industry,* 6th ISOEN Proceedings, Tübingen (Germany) **(1999)** pp. 376-378.

[4] M. Morvan, T. Talou and J.-F. Beziau, *Quality control of polyuretnane foams for automotive industry by electronic noses, gas chromatography – mass spectrometry and sensory analysis*, Pittcon 2000, New Orleans (USA) **(2000)**.

[5] S.J. Martin, V.E. Granstaff, and G.C. Frye, *Characterization of a quartz microbalance with simultaneous mass and liquid loading*, Anal. Chem., 63 **(1991)** pp. 2272-2281.

[6] C. Behling, R. Lucklum, and P. Hauptmann, *Response of quartz-crystal resonators to gas and liquid analyte exposure*, Sensors and Actuators, A 68 **(1998)** pp. 388-398.

[7] R. Lucklum, C. Behling, and P. Hauptmann, *Role of mass accumulation and viscoelastic film properties for the response of acoustic-wave-based chemical sensors*, Anal. Chem., 71 **(1999)** pp. 2488-2496.

[8] E.-L. Kalman, F. Winquist , I. Lundström, M. Grönberg, and A. Löfvendah, *A semiconductor gas sensor array for the detection of gas emissions from interior trim materials in automobiles*. JSAE 980995 **(1998)** pp. 1-6.

[9] H.J. Macfie and N. Bratchell, *Designs to balance the effect of order of presentation and first-order of carry-over effects in hall tests*. J. Sensory Studies, 4 **(1989)** pp. 129-148.

[10] J.N. Jaubert, C. Tapiero and J.C. Dore, *The field of odors : toward a universal language for odor relationships*. Perf. Flav., 20 **(1995)** pp. 1-16.

[11] S. Garrigues, PhD thesis, in preparation **(2001)**.

Acknowledgements

Authors thank HKR Sensorsysteme GmbH staff for its technical support

This research was conducted in partial fulfilment of the requirements of the doctoral thesis of Sandrine GARRIGUES.

AUTHOR INDEX

SUBJECT INDEX